Advancing Maths for AQA

DECISION MATHS I

David Pearson and Victor Bryant

eries editors
am Boardman Ted Graham David Pearson
oger Williamson

	Minimum connectors	1	
	Shortest path problem (Dijkstra's algorithm)	23	2
	Chinese postman problem	44	3
	Travelling salesman problem	65	4
	Graph theory	90	5
	Matchings	109	6
	Sorting algorithms	123	7
	Algorithms	135	8
	Linear programming	150	9
	am style practice paper	177	
	swers	181	
	dex	213	

Heinemann is an imprint of Pearson Education Limited, a company incorporated in England and Wales, having its registered office at Edinburgh Gate, Harlow, Essex, CM20 2JE. Registered company number: 872828

Heinemann is the registered trademark of Pearson Education Limited

First published 2004

08
10 9 8 7 6

British Library Cataloguing in Publication Data is available from the British Library on request.

13-digit ISBN: 978 0 435513 35 1

Edited by Lesley Montford
Typeset and illustrated by Tech-Set Limited, Gateshead, Tyne & Wear
Original illustrations © Harcourt Education Limited, 2004
Cover design by Miller, Craig and Cocking Ltd
Printed in Great Britain by Scotprint

Acknowledgements
The publishers and authors acknowledge the work of the writers Ray Atkin, John Berry, Derek Collins, Tim Cross, Ted Graham, Phil Rawlins, Tom Roper, Rob Summerson, Nigel Price, Frank Chorlton and Andy Martin of the *AEB Mathematics for AQA A-Level Series*, from which some exercises and examples have been taken.

The publishers and authors also acknowledge the work of Keith Parramore and Joan Stephens in *Decision Mathematics 1*, from which some exercises and examples have been taken.

The publishers' and authors' thanks are due to AQA for permission to reproduce questions from past examination papers.

The answers have been provided by the authors and are not the responsibility of the examining board.

Every effort has been made to contact copyright holders of material reproduced in this book. Any omissions will be rectified in subsequent printings if notice is given to the publishers.

About this book

This book is one in a series of textbooks designed to provide you with exceptional preparation for AQA's new Advanced GCE Specification B. The series authors are all senior members of the examining team and have prepared the textbooks specifically to support you in studying this course.

Finding your way around

The following are there to help you find your way around when you are studying and revising:

- **edge marks** (shown on the front page) – these help you to get to the right chapter quickly;
- **contents list** – this identifies the individual sections dealing with key syllabus concepts so that you can go straight to the areas that you are looking for;
- **index** – a number in bold type indicates where to find the main entry for that topic.

Key points

Key points are not only summarised at the end of each chapter but are also boxed and highlighted within the text like this:

> For every action, there is an equal but opposite reaction.

Exercises and exam questions

Worked examples and carefully graded questions familiarise you with the specification and bring you up to exam standard. Each book contains:

- Worked examples and Worked exam questions to show you how to tackle typical questions; Examiner's tips will also provide guidance;
- Graded exercises, gradually increasing in difficulty up to exam-level questions, which are marked by an [A];
- Test-yourself sections for each chapter so that you can check your understanding of the key aspects of that chapter and identify any sections that you should review;
- Answers to the questions are included at the end of the book.

This book has been divided into three sections.

Section 1 (Chapters 1–5) investigates networks. I recommend working through the book in the order of the chapters. Chapters 1–4 cover solving practical problems before studying graph/network theory in Chapter 5.

Section 2 (Chapters 6–8) cover algorithms. Again I suggest leaving the chapter on general algorithms to last.

Section 3 (Chapter 9) investigates linear programming, with emphasis on the graphical approach to problem solving. Although this section only has one chapter, it is certainly the biggest single topic of the book.

1 Minimum connectors

Learning objectives 1
1.1 Introduction 1
1.2 Kruskal's algorithm 3
1.3 Prim's algorithm 7
1.4 Comparison of Prim's and Kruskal's algorithms 11
1.5 Matrix form of Prim's algorithm 12
Key point summary 20
Test yourself 21

2 Shortest path problem (Dijkstra's algorithm)

Learning objectives 23
2.1 Introduction 23
2.2 Triangles in networks 24
2.3 Dijkstra's algorithm 25
2.4 Multiple start points 30
2.5 Limitations 32
2.6 Finding the route 33
Key point summary 41
Test yourself 42

3 Chinese postman problem

Learning objectives 44
3.1 Introduction 44
3.2 Traversable graphs 45
3.3 Pairing odd vertices 47
3.4 Chinese postman algorithm 48
3.5 Finding a route 50
3.6 Variations of the Chinese postman problem 52
Key point summary 62
Test yourself 63

4 Travelling salesman problem

Learning objectives 65
4.1 Introduction 65
4.2 Upper bound 67
4.3 Nearest-neighbour algorithm 68
4.4 Limitations of the nearest-neighbour algorithm 71
4.5 Lower bound 72

4.6 Incomplete networks 76
Key point summary 88
Test yourself 88

5 Graph theory
Learning objectives 90
5.1 Introduction 90
5.2 Definitions 90
Key point summary 106
Test yourself 107

6 Matchings
Learning objectives 109
6.1 Introduction 109
6.2 Matchings 112
6.3 Improving a matching to obtain a maximal matching 113
Key point summary 121
Test yourself 121

7 Sorting algorithms
Learning objectives 123
7.1 Introduction 123
7.2 Bubble sort 123
7.3 Shuttle sort 126
7.4 Shell sort 127
7.5 Quick sort 129
7.6 Efficiency of methods 131
Key point summary 133
Test yourself 133

8 Algorithms
Learning objectives 135
8.1 Introduction 135
8.2 Flow diagrams 137
8.3 Set of instructions written in pseudo English 140
8.4 Stopping conditions 142
Key point summary 148
Test yourself 148

9 Linear programming
Learning objectives 150
9.1 Introduction 150
9.2 Graphs of inequalities 151
9.3 Formulation of the problem 153
9.4 Graphical solutions 157
Key point summary 175
Test yourself 176

Exam style practice paper 177

Answers 181

Index 213

CHAPTER 1

Minimum connectors

Learning objectives

After studying this chapter, you should be able to:

- understand the significance of a minimum spanning tree
- apply Kruskal's algorithm to a network
- apply Prim's algorithm to a network
- apply Prim's algorithm to a matrix.

1.1 Introduction

A haulage company has many lorries collecting and delivering goods throughout the country. The company uses the lorries in such a way that the total distance travelled by all of the lorries is as short as possible. This is one example of a company trying to find a minimum.

The aim of television cabling companies is to connect as many homes in as many towns as possible to a network. The less this costs the better, so they aim to use as short a length of cabling as possible.

The map below shows roads connecting main towns near to Manchester.

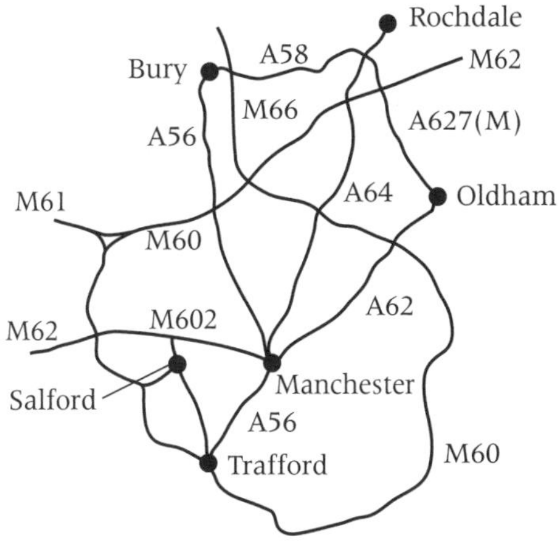

The problem is to determine how to connect six towns in the network using the least amount of cabling.

We can model this problem by drawing a **network**. The numbers on the network represent the distances in miles between towns.

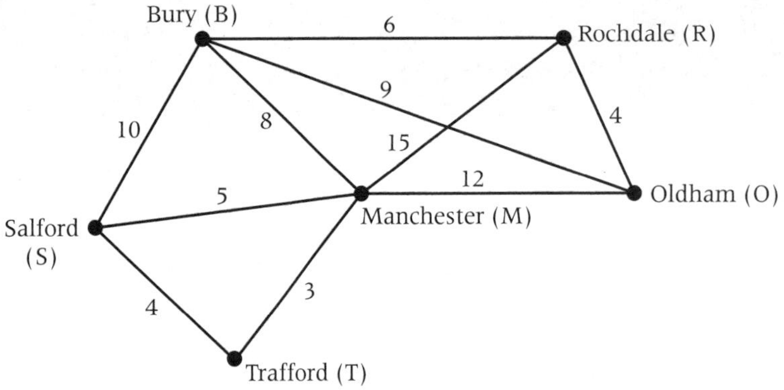

The towns are vertices (or nodes) on the network. The lines connecting the towns are called edges (or arcs).
The problem is to determine how to connect the vertices R, O, M, T, S and B so that the total length of the edges is as short as possible.

The solution can be found by inspection. The solution is 25 miles of cabling, as shown below.

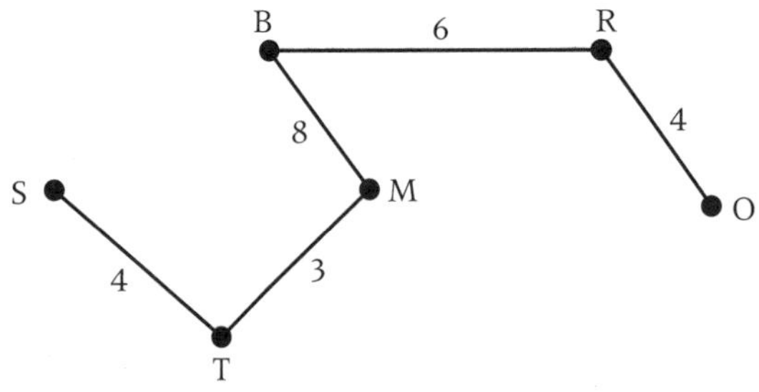

It is interesting to notice that the solution contains five edges. The number of edges in a minimum spanning tree is always one less than the number of vertices. The edge MS of length 5 was not included as this edge would make a **cycle** MSTM. (Graph theory is dealt with fully in Chapter 5.)

> This is called the **minimum spanning tree** for the network.

In this example there was only one spanning tree that gave the minimum answer of 25, but this is not always the case.
It was easy to find a minimum spanning tree by inspection in this example because there were only six towns.
However, if there were 20 towns then it would be impractical to try and find a minimum spanning tree by inspection as the number of possible spanning trees for a network of 20 vertices is 2.6×10^{23}.
To find a solution efficiently we apply an algorithm.

Algorithms are dealt with in Chapter 8.

There are two algorithms for finding a minimum spanning tree: Kruskal's and Prim's.

1

1.2 Kruskal's algorithm

To find a minimum spanning tree for a network with n edges.

Step 1 Choose the unused edge with the lowest value.

Step 2 Add this edge to your tree.

Step 3 If there are $n - 1$ edges in your tree, stop. If not, go to Step 1.

At each step remember to make sure you do not make a cycle.

Worked example 1.1

Use Kruskal's algorithm to find the minimum spanning tree for the following network.

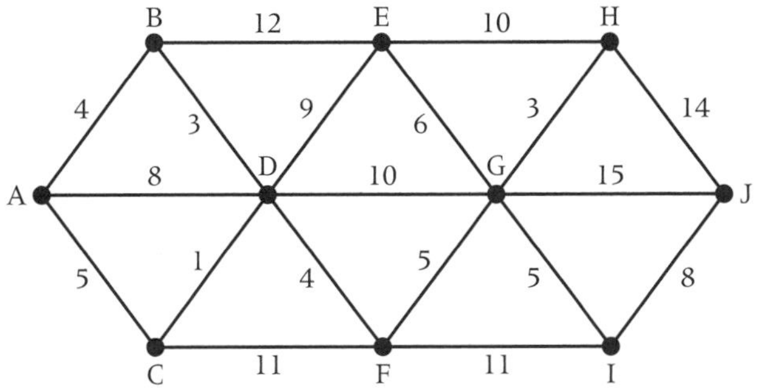

Solution

There are 10 vertices so the minimum spanning tree will have nine edges.

Step 1 Choose the lowest edge CD, a value of 1.

Step 2 Add CD to the tree.

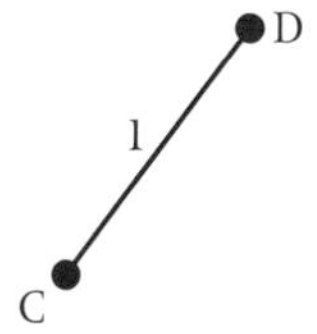

Step 1 Choose the lowest edge BD or GH, both have a value of 3. It doesn't matter which is chosen.

Step 2 Add BD to the tree.

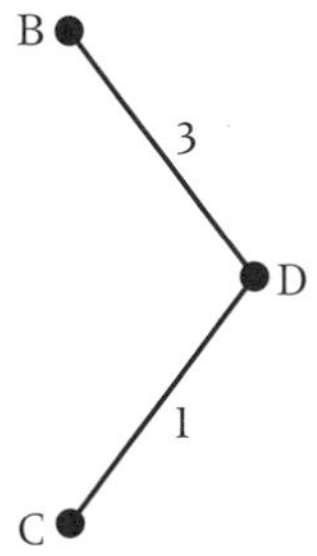

Step 1 Choose the lowest edge GH, with a value of 3.

Step 2 Add GH to the tree.

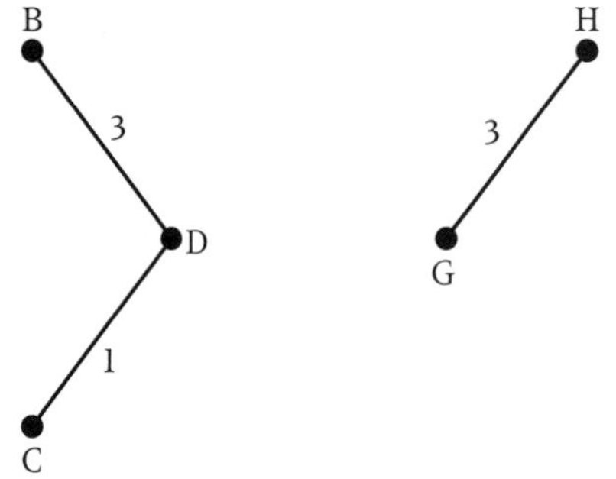

Step 1 Choose the lowest edge AB or DF, both have a value of 4.

Step 2 Add AB to the tree.

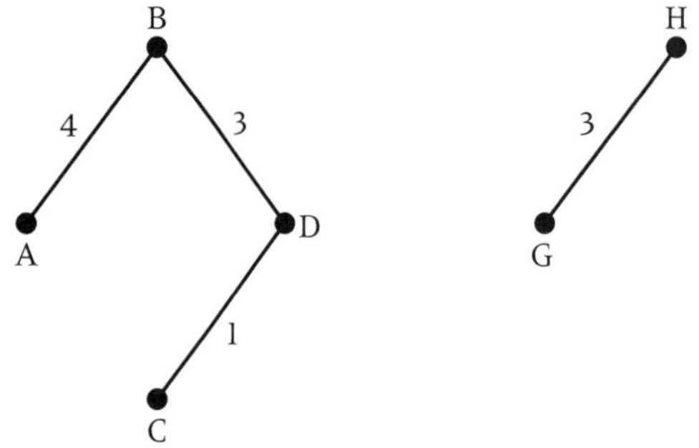

Step 1 Choose the lowest edge DF, with a value of 4.

Step 2 Add DF to the tree.

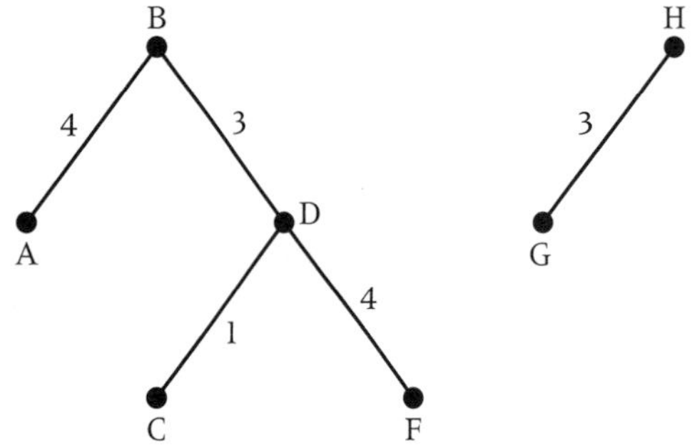

Step 1 Choose the lowest edge AC, FG or GI, all have a value of 5.

Step 2 **DO NOT ADD** AC to the tree. This would make a cycle of ABDCA. Add FG to the tree.

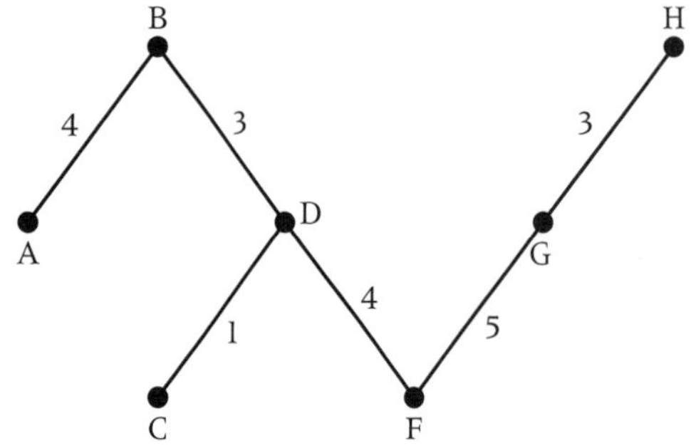

Step 1 Choose the lowest edge GI, with a value of 5.

Step 2 Add GI to the tree.

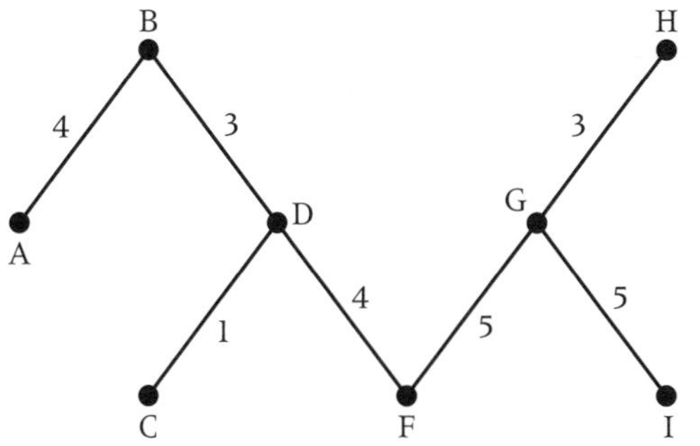

Step 1 Choose the lowest edge EG, with a value of 6.

Step 2 Add EG to the tree.

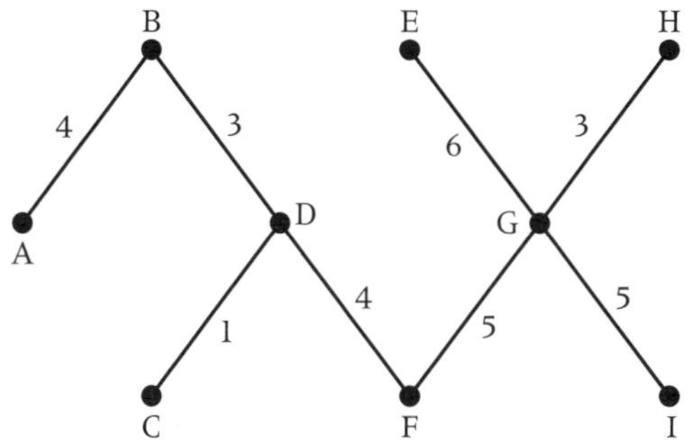

Step 1 Choose the lowest edge AD or IJ, both with a value of 8.

Step 2 **DO NOT ADD** AD to the tree as this would make a cycle of ABDA. Add IJ to the tree.

Step 3 There are now nine edges and the minimum spanning tree is complete.

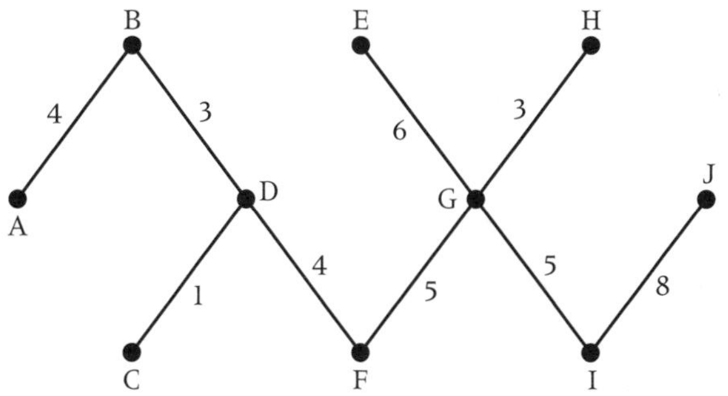

The length of the minimum spanning tree is 39 units.

EXERCISE 1A

1 Showing your working, use Kruskal's algorithm on the following networks to find a minimum spanning tree. In each case, state the length of your minimum spanning tree and draw it.

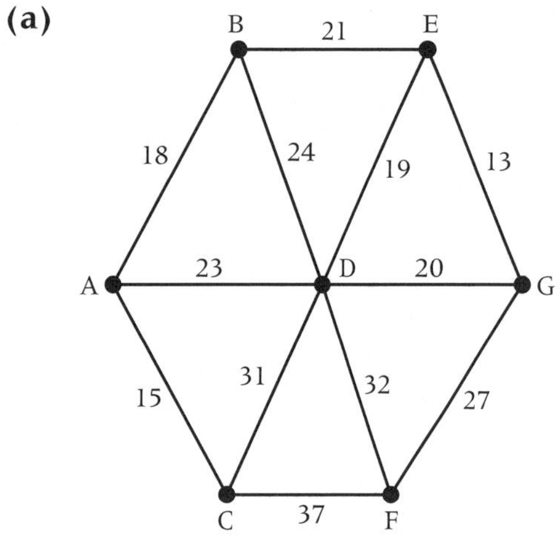

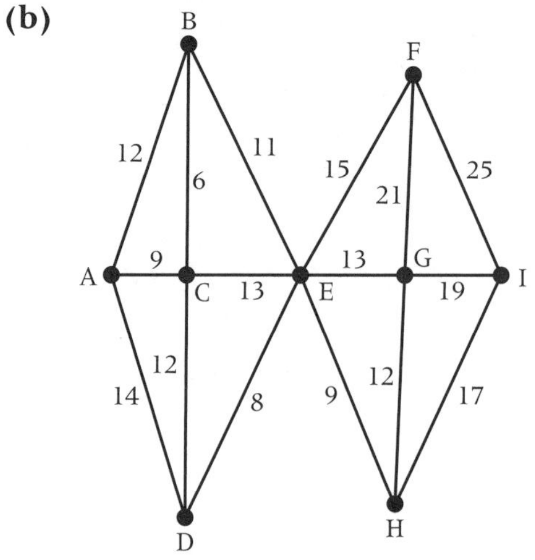

2 The diagram below shows a network of roads connecting villages.

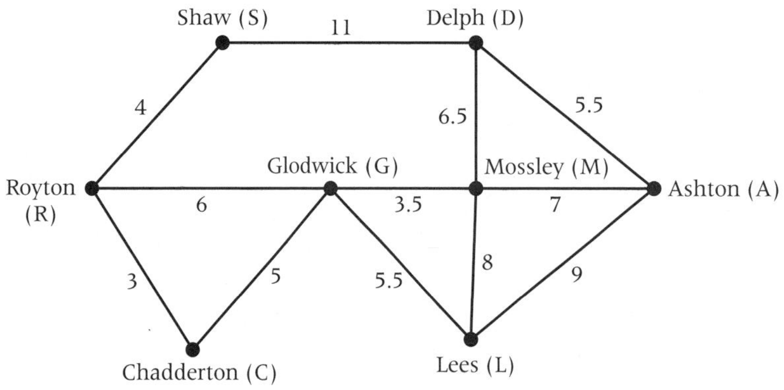

Showing your working, use Kruskal's algorithm to find a minimum spanning tree. State the length of your minimum spanning tree and draw it.

3 The diagram below shows rooms in a school. The numbers on each edge represent the length of cabling required to connect that particular arc to a new cabling network. Each room is to be connected to this new computer network.

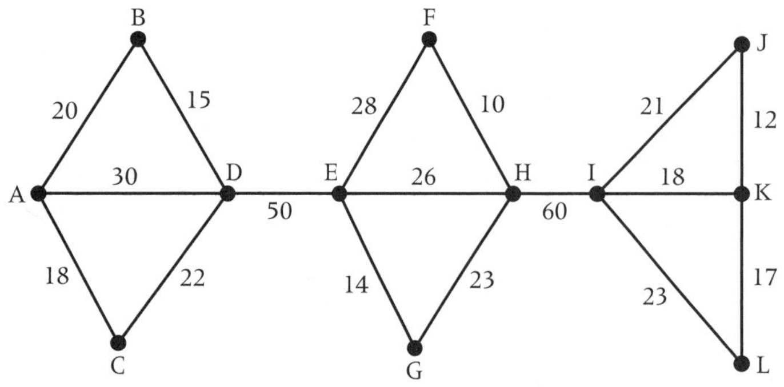

Showing your working, use Kruskal's algorithm to find a minimum spanning tree. State the length of your minimum spanning tree and draw it.

1.3 Prim's algorithm

To find a minimum spanning tree for a network with n edges.

Step 1 From a start vertex draw the lowest valued edge to start your tree. (Any vertex can be chosen as the start vertex; however, it will always be given in an exam question.)

Step 2 From any vertex on your tree, add the edge with the lowest value.

Step 3 If there $n - 1$ edges in your tree, you have finished. If not, go to step 2.

At each step remember to make sure you do not make a cycle.

A key aspect of Prim's algorithm is that as the tree is being built, the tree is always connected. You are always adding a new vertex to your current tree.

Worked example 1.2

Use Prim's algorithm, starting from A, to find the minimum spanning tree for the network below.

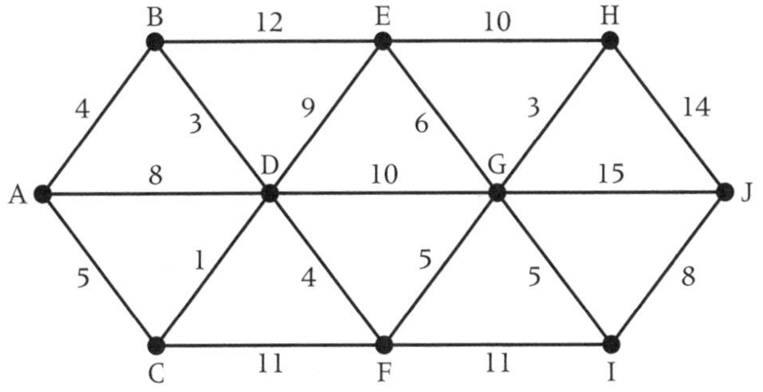

Solution

Step 1 Choose the lowest edge from A, which is AB, with a value of 4.

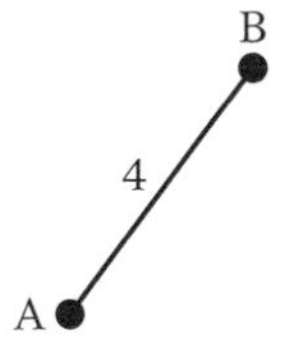

Step 2 Choose the lowest edge from A or B, which is BD, with a value of 3. Add BD to the tree.

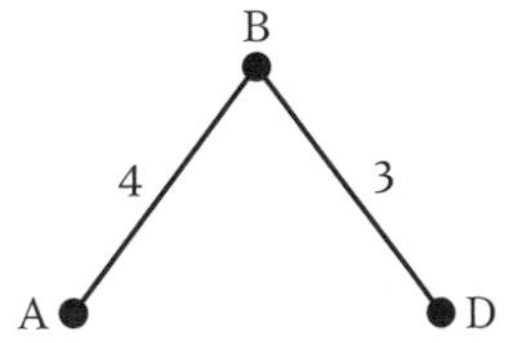

Step 2 Choose the lowest edge from A, B or D, which is DC, with a value of 1. Add DC to the tree.

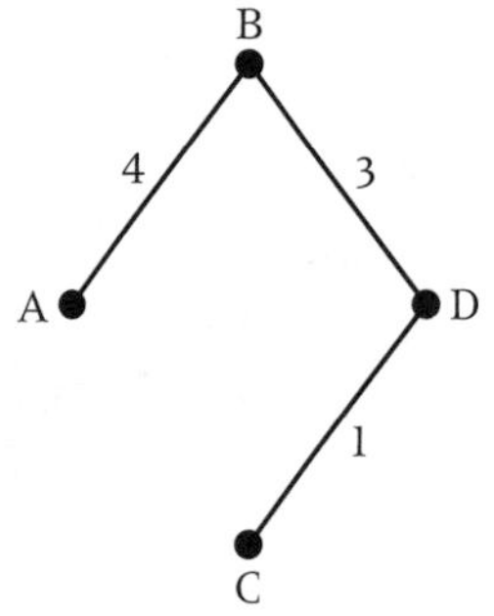

Step 2 Choose the lowest edge from A, B, C or D, which is DF, with a value of 4. Add DF to the tree.

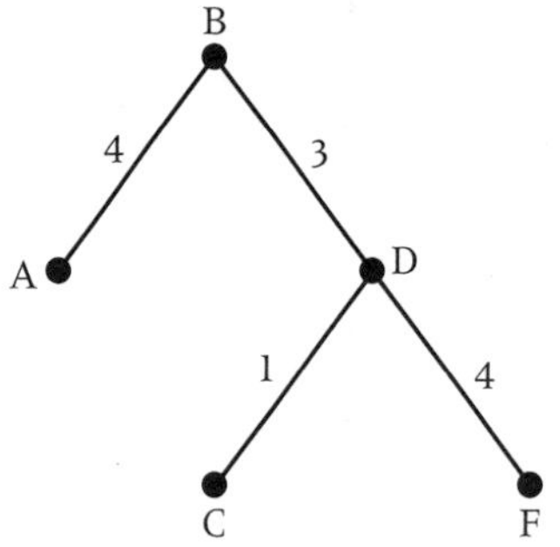

Step 2 Choose the lowest edge from A, B, C, D or F, which is either AC or FG, with a value of 5. **DO NOT ADD** AC to the tree as this would make a cycle of ADCA. Add FG to the tree.

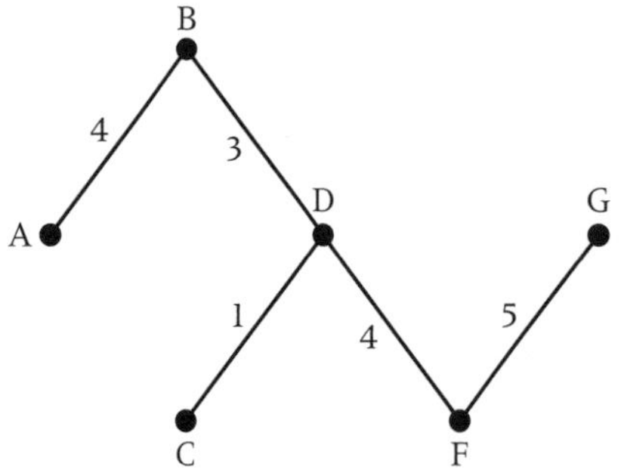

Step 2 Choose the lowest edge from A, B, C, D, F or G, which is GH, with a value of 3. Add GH to the tree.

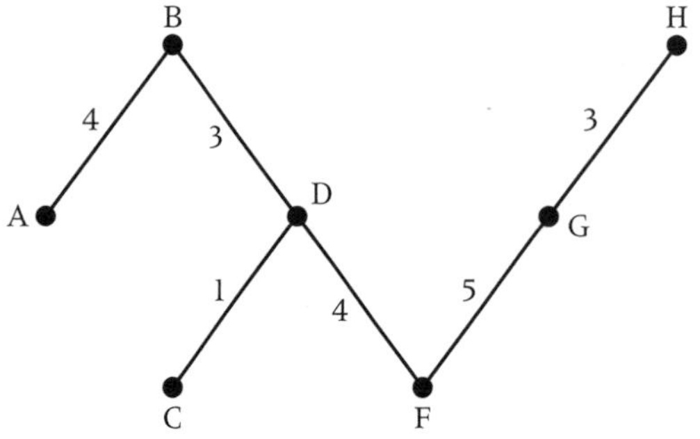

Step 2 Choose the lowest edge from A, B, C, D, F, G or H, which is GI, with a value of 5. Add GI to the tree.

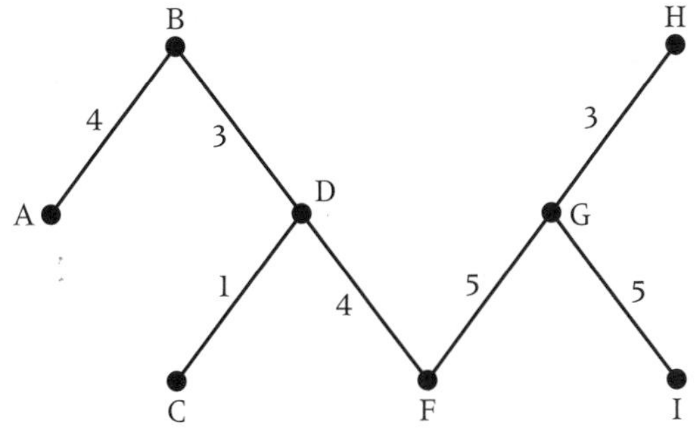

Step 2 Choose the lowest edge from A, B, C, D, F, G, H or I, which is GE, with a value of 6. Add GE to the tree.

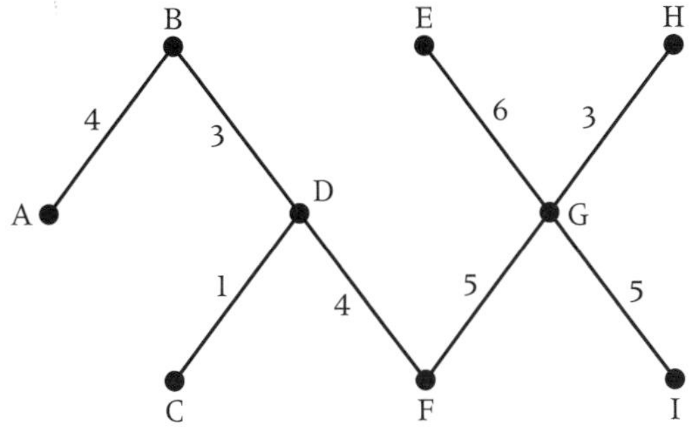

Step 2 Choose the lowest edge from A, B, C, D, F, G, H, I or E, which is either AD or IJ, with a value of 8. **DO NOT CHOOSE** AD, as this would make a cycle of ABDA. Add IJ to the tree.

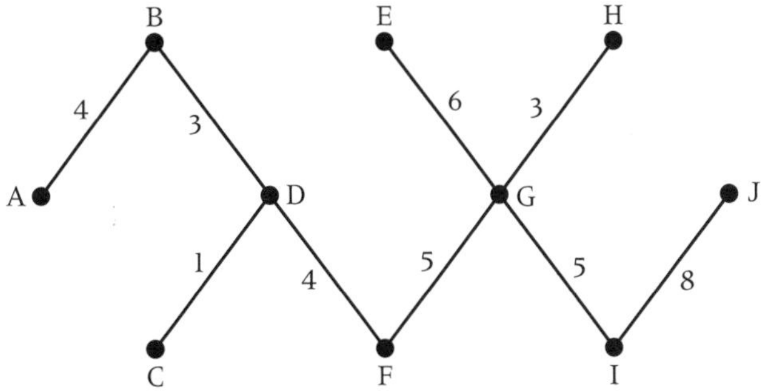

Step 3 There are now $n - 1$ (which equals 9, in this case) edges. The minimum spanning tree is your final answer.

The length of the minimum spanning tree is 39 units.

EXERCISE 1B

1 Showing your working, use Prim's algorithm starting from A on the following networks to find a minimum spanning tree. In each case, state the length of your minimum spanning tree and draw it.

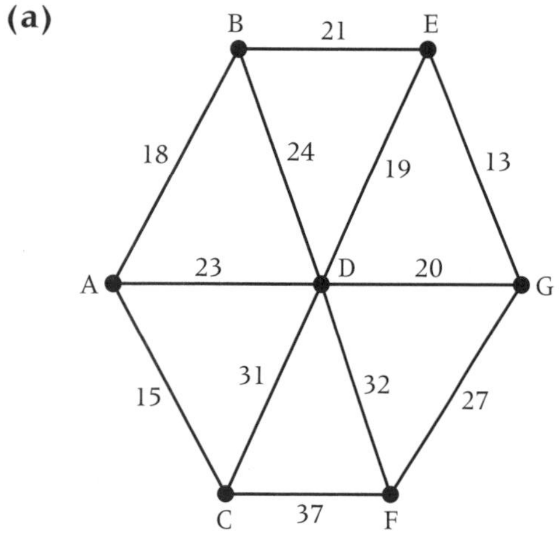

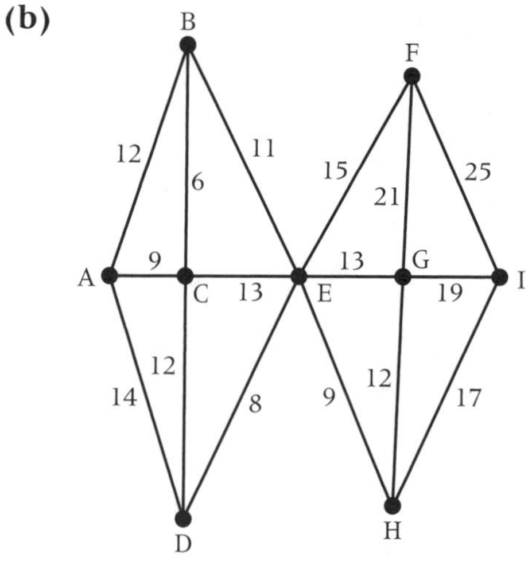

2 The diagram below shows a network of roads connecting villages.

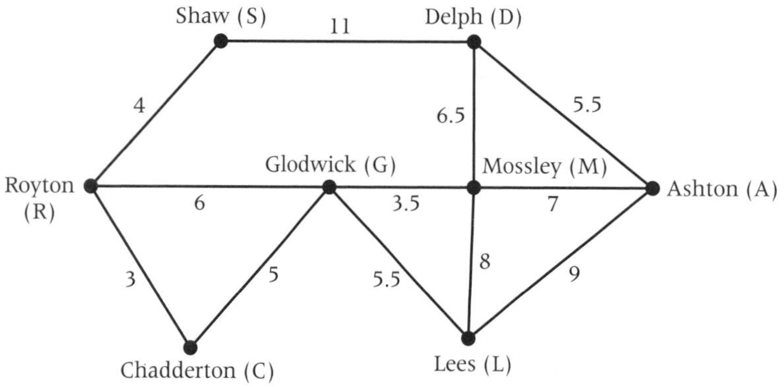

Showing your working, use Prim's algorithm starting from A to find a minimum spanning tree. State the length of your minimum spanning tree and draw it.

3 The diagram below shows rooms in a school. The numbers on each edge represent the length of cabling required to connect that particular arc to a new cabling network. Each room is to be connected to this new computer network.

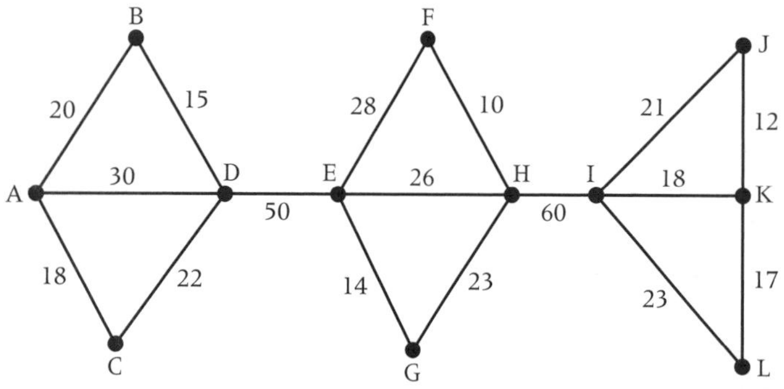

Showing your working, use Prim's algorithm starting from A to find a minimum spanning tree. State the length of your minimum spanning tree and draw it.

1.4 Comparison of Prim's and Kruskal's algorithms

Whenever these two algorithms are applied to the same network, the order in which we add the edges is different. However, the final minimum spanning tree will always be the same.

It is essential that you **write down** the order that each edge is added to the tree. (You do not have to draw the tree at each step of the algorithm.) In an examination this is how an examiner will determine if the correct algorithm has been used.

Worked example 1.3

Find the minimum spanning tree for the network below, showing the order in which edges are selected:

(a) using Kruskal's algorithm,

(b) using Prim's algorithm starting from A.

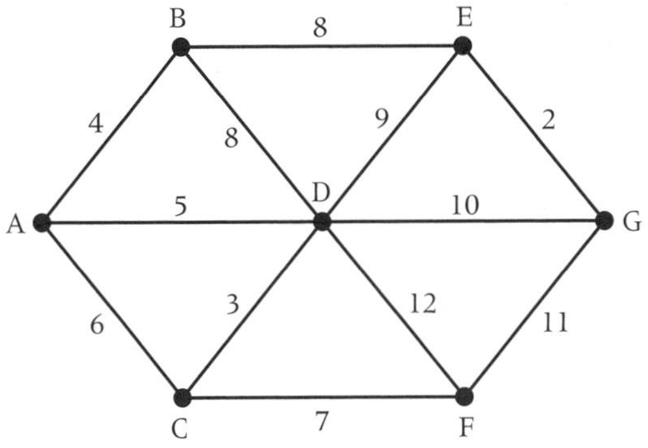

Solution

The minimum spanning tree for the network is:

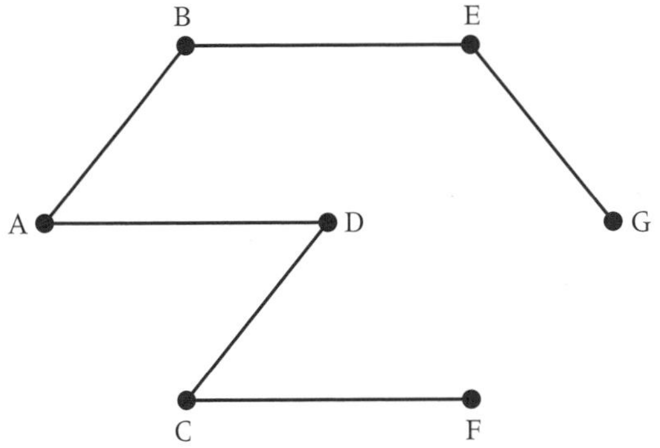

The length of the minimum spanning tree is 29.

The order of adding the edges to the tree is different using the two algorithms:

(a) Kruskal's	(b) Prim's
EG 2	AB 4
CD 3	AD 5
AB 4	DC 3
AD 5	CF 7
CF 7	BE 8
BE 8	EG 2

1.5 Matrix form of Prim's algorithm

Often maps contain tables that summarise distances between towns to enable a reader to find the distance between two places at a glance.

A table can be called a **matrix**.

The table below shows the distance, in miles, between five cities.

	Aberdeen	Birmingham	Cardiff	London	Oxford
Aberdeen	–	431	531	544	503
Birmingham	431	–	109	120	68
Cardiff	531	109	–	152	105
London	544	120	152	–	56
Oxford	503	68	105	56	–

How can we find a minimum spanning tree when information is given in this format?

From the table it is possible to draw the network connecting the towns, and then from this network to apply either of the algorithms. This method is time-consuming for five towns and certainly would be impractical if there were 20 towns.
Prim's algorithm can be adapted to enable a minimum spanning tree to be determined directly from the table (matrix).

Step 1 Label the column corresponding to the start vertex with a 1. Delete the row corresponding to that vertex.

Step 2 Ring the smallest available value in any labelled column.

DO NOT CONSIDER ENTRIES THAT HAVE BEEN DELETED

Step 3 Label the column corresponding to the ringed vertex with a 2, etc. Delete the row corresponding to that vertex.

Step 4 Repeat steps 2 and 3 until all rows have been deleted.

Step 5 Write down the order in which edges were selected and the length of the minimum spanning tree.

Worked example 1.4

Use Prim's algorithm, starting from Aberdeen, to find the minimum spanning tree for the following network represented in matrix form. Draw the minimum spanning tree.

	Aberdeen	Birmingham	Cardiff	London	Oxford
Aberdeen	–	431	531	544	503
Birmingham	431	–	109	120	68
Cardiff	531	109	–	152	105
London	544	120	152	–	56
Oxford	503	68	105	56	–

Solution

Step 1 Label the Aberdeen column with a 1 and delete the row corresponding to Aberdeen.

Step 2 Ring the smallest entry, 431, in this column.

Step 3 Label the Birmingham column with a 2 and delete the row corresponding to Birmingham.

	Aberdeen①	Birmingham②	Cardiff	London	Oxford
~~Aberdeen~~	–	431	531	544	503
~~Birmingham~~	431		109	120	68
Cardiff	531	109	–	152	105
London	544	120	152	–	56
Oxford	503	68	105	56	–

Step 2 The smallest entry in either the Aberdeen or Birmingham column is 68, Birmingham to Oxford. Ring 68.

Step 3 Label the Oxford column with a 3 and delete the Oxford row.

	Aberdeen①	Birmingham②	Cardiff	London	Oxford③
~~Aberdeen~~	–	431	531	544	503
~~Birmingham~~	431		109	120	68
Cardiff	531	109	–	152	105
London	544	120	152	–	56
~~Oxford~~	503	68	105	56	

Step 2 The smallest entry in columns Aberdeen, Birmingham or Oxford is 56, Oxford to London. Ring 56.

Step 3 Label the London column with a 4 and delete the London row.

	Aberdeen①	Birmingham②	Cardiff	London④	Oxford③
~~Aberdeen~~	–	431	531	544	503
~~Birmingham~~	431		109	120	68
Cardiff	531	109	–	152	105
~~London~~	544	120	152		56
~~Oxford~~	503	68	105	56	

Step 2 The smallest entry in any of the four columns is 105, Oxford to Cardiff. Ring 105.

Step 3 Label the Cardiff column with a 5 and delete the Cardiff row.

Step 4 All rows have now been deleted.

	Aberdeen ①	Birmingham ②	Cardiff ⑤	London ④	Oxford ③
~~Aberdeen~~	~~–~~	~~431~~	~~531~~	~~544~~	~~503~~
~~Birmingham~~	~~(431)~~		~~109~~	~~120~~	~~68~~
~~Cardiff~~	~~531~~	~~109~~		~~152~~	~~(105)~~
~~London~~	~~544~~	~~120~~	~~152~~		~~(56)~~
~~Oxford~~	~~503~~	~~(68)~~	~~105~~	~~56~~	

Step 5 The order of the edges selected was AB, BO, OL and OC and the total length is 660 miles.

EXERCISE 1C

1 Showing your working, use Prim's algorithm starting from A on the following matrices to find a minimum spanning tree. In each case, state the length of your minimum spanning tree and draw it.

(a)

	A	B	C	D	E
A	–	12	7	15	9
B	12	–	8	11	14
C	7	8	–	6	17
D	15	11	6	–	13
E	9	14	17	13	–

(b)

	A	B	C	D	E	F
A	–	37	15	29	42	14
B	37	–	24	16	12	28
C	15	24	–	21	38	11
D	29	16	21	–	25	31
E	42	12	38	25	–	33
F	14	28	11	31	33	–

2 Write the network given in Worked Example 1.1 as a matrix. Apply Prim's algorithm to this matrix, starting from A, to find the minimum spanning tree.

3 Write the networks given in Exercise 1A as a matrix. Apply Prim's algorithm to this matrix, starting from A, to find the minimum spanning tree.

MIXED EXERCISE

1 The following matrix shows the distances, in miles, between six towns.

	A	B	C	D	E	F
A	–	12	20	14	15	25
B	12	–	23	7	13	8
C	20	23	–	16	16	12
D	14	7	16	–	20	9
E	15	13	16	20	–	15
F	25	8	12	9	15	–

(a) Using Kruskal's algorithm and showing your working at each stage, find the minimum spanning tree for these six towns.

(b) State the length of your minimum spanning tree. [A]

2 (a) For a connected graph with n vertices, state the number of edges in a minimum spanning tree.

(b) A cable company has laid cables to Accrington and is now considering extending the network to neighbouring villages. The distances, in miles, between the villages are shown below.

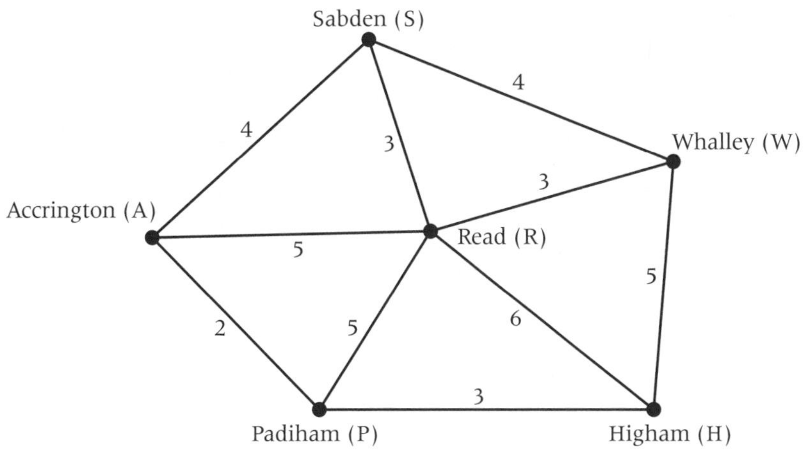

(i) Using Prim's algorithm, starting from Accrington, showing a sketch at each stage, obtain the minimum spanning tree for the cable company.

(ii) State the length of your minimum spanning tree. [A]

3 The distances, in miles, between six towns are given by the following matrix.

	A	B	C	D	E	F
A	–	3	5	10	13	19
B	3	–	4	7	13	23
C	5	4	–	7	10	22
D	10	7	7	–	17	18
E	13	13	10	17	–	9
F	19	23	22	18	9	–

(a) Using Prim's algorithm and showing your working at each stage, find a minimum spanning tree for these six towns.

(b) State the length of your minimum spanning tree.

(c) When information is provided in matrix form, explain why Prim's algorithm, in preference to Kruskal's algorithm, is normally used to find a minimum spanning tree. [A]

4 The distance chart below gives the distances, in miles, between six towns in northern England.

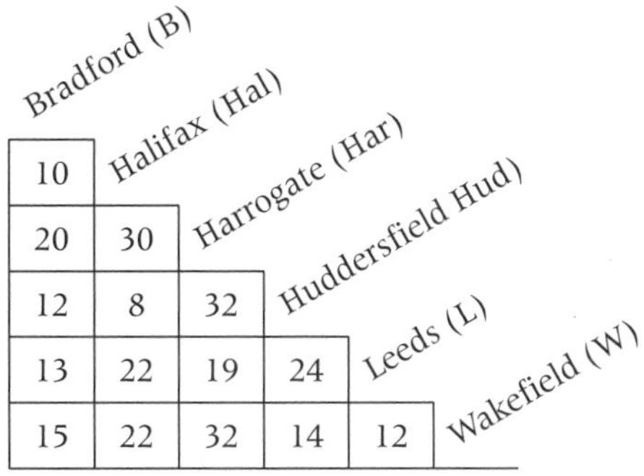
Bradford (B)
Halifax (Hal)
Harrogate (Har)
Huddersfield Hud)
Leeds (L)
Wakefield (W)
10
20 30
12 8 32
13 22 19 24
15 22 32 14 12

(a) Using Kruskal's algorithm and showing your working at each stage, find a minimum spanning tree for these six towns.

(b) State the length of your minimum spanning tree. [A]

5 The following diagram shows a network of roads connecting eight towns. The number on each arc represents the distance, in miles, between two towns.

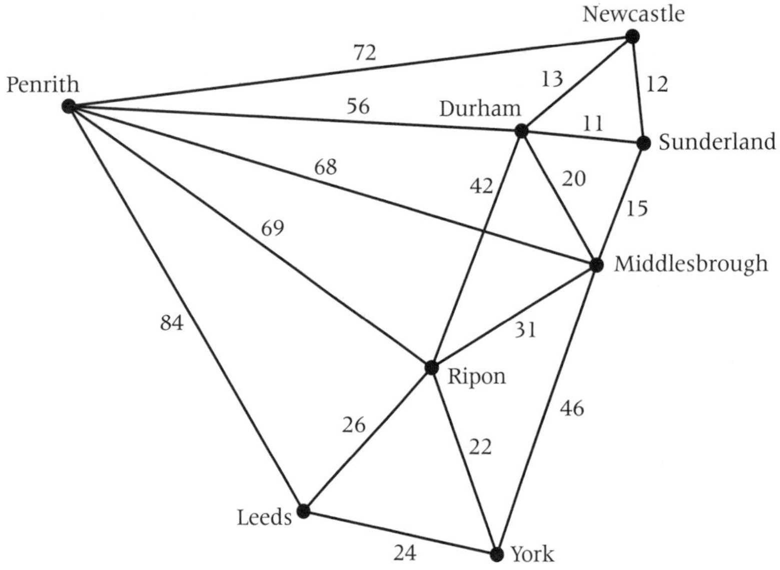

(a) Starting from Ripon and showing your working at each stage, use Prim's algorithm to find the minimum spanning tree for the eight towns. State the length of your minimum spanning tree.

(b) Draw your minimum spanning tree. [A]

6 The following diagram shows the lengths of roads, in miles, connecting eight towns.

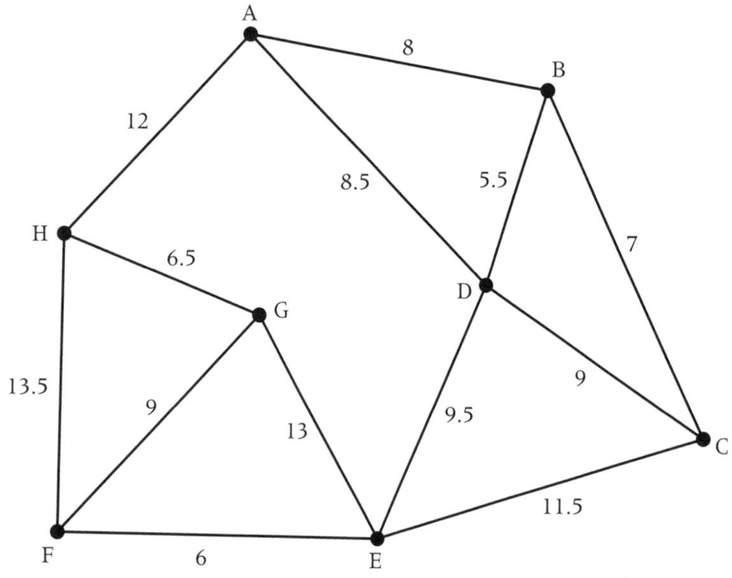

Using Kruskal's algorithm, showing your working at each stage, find the minimum spanning tree for the network. State its length. [A]

7 The following diagram shows the lengths, in miles, of roads connecting 10 towns.

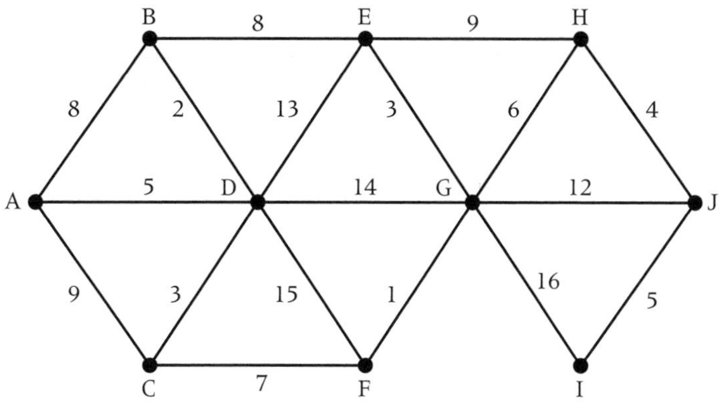

Use Kruskal's algorithm, showing the order in which you select the edges, to find the minimum spanning tree for the network. Draw your minimum spanning tree and state its length.

8 The following matrix shows the costs of connecting together each possible pair from six computer terminals:

	A	B	C	D	E	F
A	–	120	200	140	135	250
B	120	–	230	75	130	80
C	200	230	–	160	160	120
D	140	75	160	–	200	85
E	135	130	160	200	–	150
F	250	80	120	85	150	–

The computers are to be connected together so that, for any pair of computers, there should be either a direct link between them or a link via one or more other computers. Use an appropriate algorithm to find the cheapest way of connecting these computers. Show your result on a network and give the total cost.

Ensure that you show clearly the steps of the algorithm. Stating the correct answer without showing how you achieved it is not sufficient. [A]

9 The network in the diagram below indicates the main road system between 10 towns and cities in the north of England.

A computer company wishes to install a computer network between these places, using cables laid alongside the roads and designed so that all places are connected (either directly or indirectly) to the main computer located at Manchester. Find the network which uses a minimum total length of cable. [A]

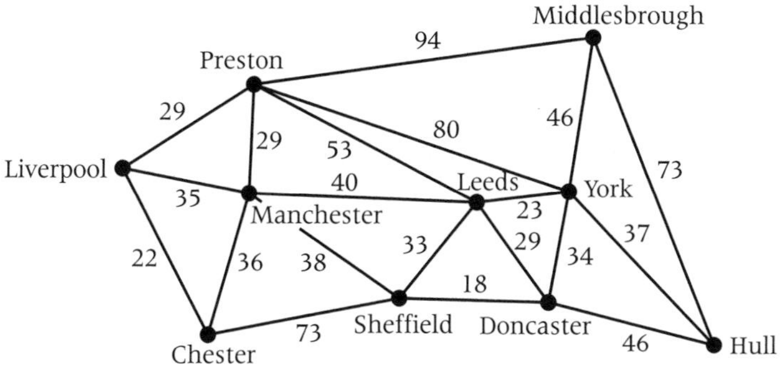

10 The table gives the distances (in miles) between six places in Scotland. Draw the network and use the graphical form of Prim's algorithm to find a minimum connector. [A]

	Aberdeen	Edinburgh	Fort William	Glasgow	Inverness	Perth
Aberdeen	–	125	147	142	104	81
Edinburgh	125	–	132	42	157	45
Fort William	147	132	–	102	66	105
Glasgow	142	42	102	–	168	61
Inverness	104	157	66	168	–	112
Perth	81	45	105	61	112	–

11 The table gives the distances (in miles) between six places in Ireland. Use the matrix form of Prim's algorithm to find a minimum connector for these places. [A]

	Athlone	Dublin	Galway	Limerick	Sligo	Wexford
Athlone	–	78	56	73	71	114
Dublin	78	–	132	121	135	96
Galway	56	132	–	64	80	154
Limerick	73	121	64	–	144	116
Sligo	71	135	80	144	–	185
Wexford	114	96	154	116	185	–

12 Some of the eight towns A to H are directly linked by roads. The table shows the distances along these roads in miles.

	A	B	C	D	E	F	G	H
A	–	10	9	5	7	6	7	–
B	10	–	7	8	–	–	8	9
C	9	7	–	–	7	8	–	8
D	5	8	–	–	–	9	9	–
E	7	–	7	–	–	8	9	7
F	6	–	8	9	8	–	6	9
G	7	8	–	9	9	6	–	–
H	–	9	8	–	7	9	–	–

(a) In the winter the local authority grits some of these roads. They wish to grit the minimum total length of roads possible in order to be able to travel between any two of the towns on gritted roads. Use Prim's algorithm starting at A to find which roads they should grit and state the minimum total length of roads which need to be gritted.

(b) Illustrate in a graph the minimum connector which you found in **(a)**. How far is it from A to B in your graph?

(c) The ambulance station is at A and next year the local authority has decided that it would like to grit the minimum total length of roads possible so that the ambulance can reach any of the other towns on gritted roads by travelling less than 15 miles. Show how to adapt your graph in **(b)** to find the roads which the local authority should grit next year. [A]

13 A complicated business document, currently written in English, is to be translated into eight of the other European Union languages. Because it is harder to find translators for some languages than for others, some translations are more expensive than others; the costs in euros are as shown in the table.

From/To	Dan	Dut	Eng	Fre	Ger	Gre	Ita	Por	Spa
Danish (Da)	–	90	100	120	60	160	120	140	120
Dutch (Du)	90	–	70	80	50	130	90	120	80
English (E)	100	70	–	50	60	150	110	150	90
French (F)	120	80	50	–	70	120	70	100	60
German (G)	60	50	60	70	–	120	80	130	80
Greek (Gr)	160	130	150	120	120	–	100	170	150
Italian (I)	120	90	110	70	80	100	–	110	70
Portuguese (P)	140	120	150	100	130	170	110	–	50
Spanish (S)	120	80	90	60	80	150	70	50	–

Use Prim's algorithm to decide which translations should be made so as to obtain a version in each language at minimum total cost. [A]

14 An oil company has eight oil rigs producing oil from beneath the North Sea, and has to bring the oil through pipes to a terminal on shore. Sometimes oil can be piped from one rig to another, and rather than build a separate pipe from each rig to the terminal the company plans to build the pipes in such a way as to minimise their total length.

Where it is possible to build pipes from one rig to another the distances (in km) are given in the table below.

	T	A	B	C	D	E	F	G	H
T	–	120	150	–	120	100	–	70	180
A	120	–	60	60	90	–	210	160	40
B	150	60	–	20	–	180	170	–	50
C	–	60	20	–	40	160	150	140	60
D	120	90	–	40	–	130	–	110	–
E	100	–	180	160	130	–	–	30	–
F	–	210	170	150	–	–	–	150	200
G	70	160	–	140	110	30	150	–	200
H	180	40	50	60	–	–	200	200	–

(a) Find the minimum length of pipes needed to connect the entire system.

(b) Find the minimum length of pipes needed to connect the terminal (T) to the oil rig F. [A]

15 An international organisation has offices A, B, C, D, E, F, G and H. The table shows the cost, in pounds, of transmitting a piece of information from one of the offices to another along all existing direct links.

From \ To	A	B	C	D	E	F	G	H
A	–	12	8	9	–	–	–	–
B	12	–	6	–	10	–	–	–
C	8	6	–	10	–	10	–	–
D	9	–	10	–	–	–	12	–
E	–	10	–	–	–	11	–	20
F	–	–	10	–	11	–	18	18
G	–	–	–	12	–	18	–	14
H	–	–	–	–	20	18	14	–

(a) Construct a network using vertices A to H to represent the information in the table.

(b) A piece of information has to be passed from office A to all the other offices, either directly or by being passed on from office to office. Use Kruskal's algorithm to find the minimum total cost of passing the information to all the offices.

(c) Office C joins a communications discount scheme in which the cost of passing information from C is halved (but it does not affect the cost of C's incoming information).

(i) Describe how to adapt the above table to show the revised costs.

(ii) Adapt your answer to **(b)** in order to calculate the minimum cost of passing a piece of information from office A to all the other offices with these revised costs. [A]

Key point summary

1 A **minimum spanning tree** shows the shortest set of edges connecting a number of vertices. There is always one edge less than the number of vertices in a spanning tree. *p2*

2 **Kruskal's algorithm** adds edges to a tree in order of size. *p3*

3 **Prim's algorithm** adds the nearest vertex to a current tree. *p7*

4 Prim's algorithm can be applied to a **matrix**. *p12*

Test yourself

What to review

1 Draw all possible spanning trees for the graph shown below. *Section 1.1*

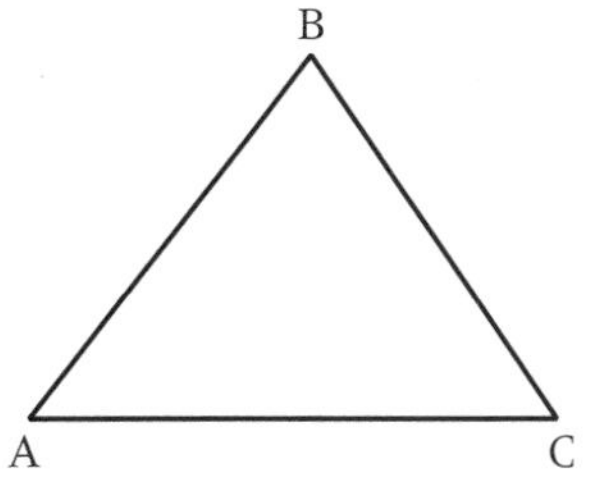

2 Find for the following network the minimum spanning tree: *Sections 1.1, 1.2, 1.3*

(a) using Prim's algorithm starting from A,

(b) using Kruskal's algorithm.

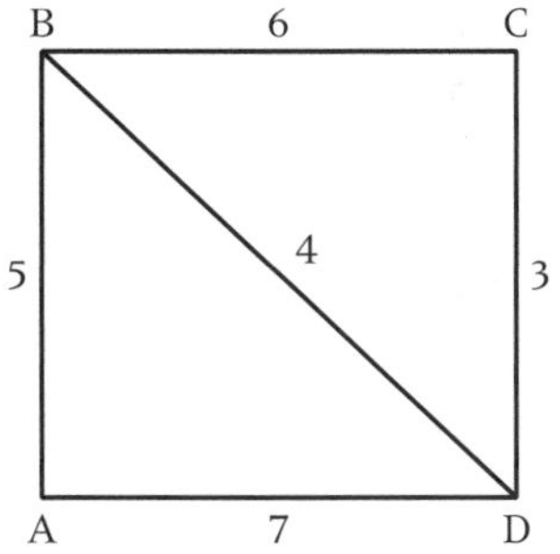

In each case list the order in which the edges are added to your spanning tree.

3 Find a minimum spanning tree for the following network, starting from A. *Section 1.5*

	A	**B**	**C**	**D**	**E**
A	–	8	12	6	11
B	8	–	9	10	7
C	12	9	–	13	14
D	6	10	13	–	5
E	11	7	14	5	–

4 Draw a graph with four vertices that have edges of length 5, 6, 7, 8, 9 and a minimum spanning tree of length 19 units. *Section 1.1*

Test yourself ANSWERS

1

2 **(a)** AB, BD, CD
Total 12

(b) CD, BD, AB

3 AD, DE, BE, BC
Total 27

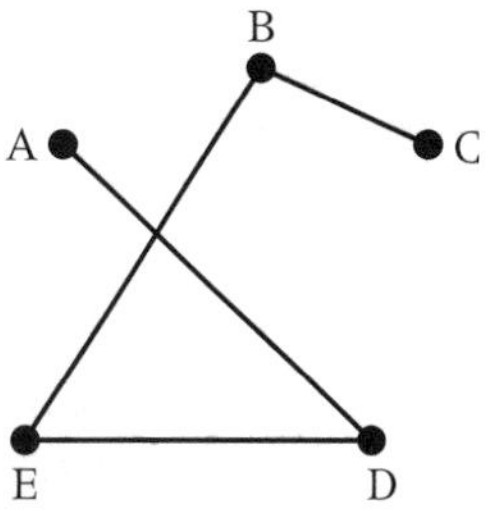

4

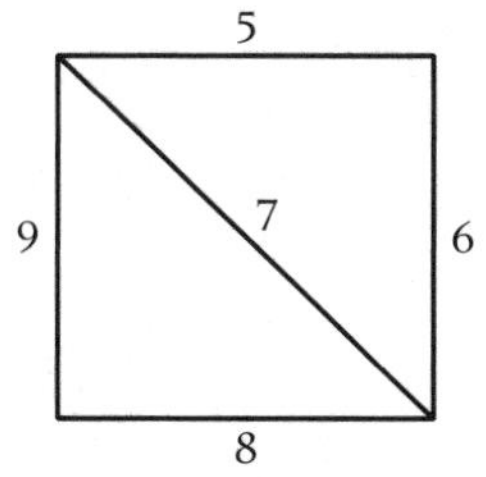

CHAPTER 2

Shortest path problem (Dijkstra's algorithm)

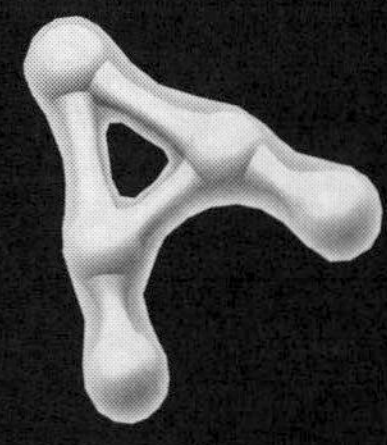

Learning objectives

After studying this chapter, you should be able to:

- apply Dijkstra's algorithm to a network
- trace back through a network to find a route corresponding to a shortest path
- apply Dijkstra's algorithm to a network with multiple start points
- understand the situations where Dijkstra's algorithm fails.

2.1 Introduction

When you plan a journey, there are different factors you might consider:

- do you want to go the shortest distance?
- do you want to take the minimum time?
- do you want to minimise the cost?

Autoroute is a computerised route planner. You enter the start and finish points of a journey and the program works out the different routes, depending on the criterion that is to be minimised.
Another innovation used today is satellite navigation systems. These keep a motorist in constant contact with a central computer. The central computer monitors traffic flow and updates the motorist on the optimum route from his or her current position. These systems are invaluable for haulage and coach companies.
How do these systems work?
The basic principle is to find the shortest route from one part of the diagram to another, in this case from A to F.

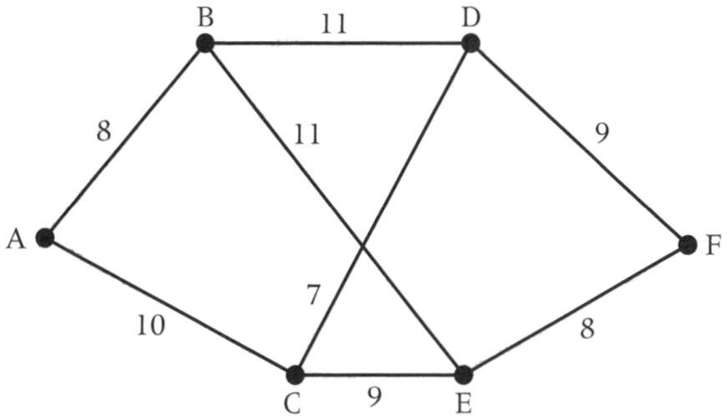

There are four routes from A to F:

ABDF, length 28; ABEF, length 27; ACDF, length 26;
ACEF, length 27

The shortest route is ACDF.

In this example there are only four possibilities to consider, but if the network were more complex then this method, called a complete enumeration, would become impractical. This chapter uses an algorithm to find the shortest path.

2.2 Triangles in networks

Consider a real-life situation in which we wish to travel from Royton to London.

The diagrams below show roads connecting Royton, Birmingham and London.

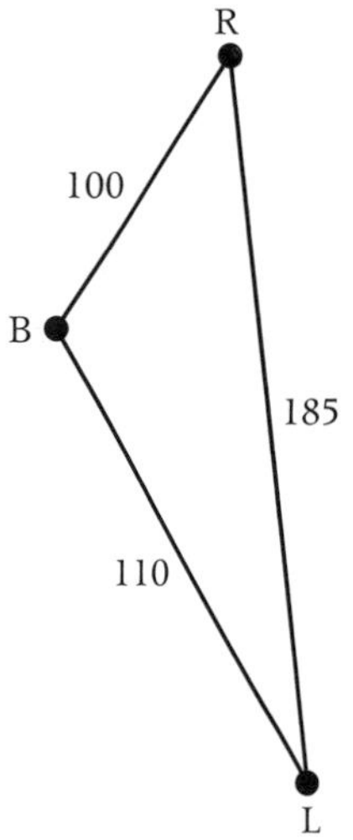

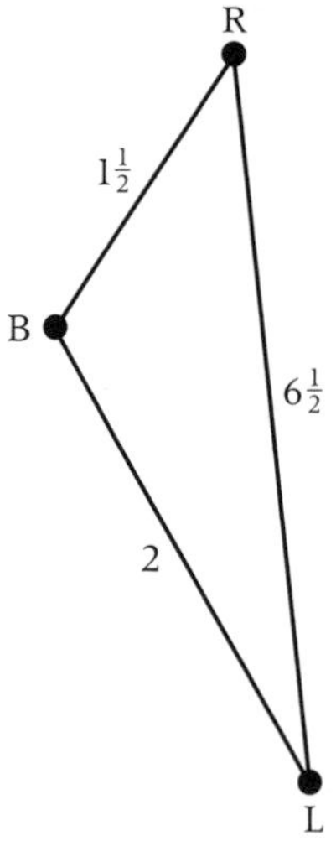

The numbers on the first diagram show the distances, in miles, between the towns.

There are two possible routes: Royton direct to London, which is 185 miles, or Royton to Birmingham and then Birmingham to London, a total distance of 210 miles.

The right-hand diagram shows the times of the same journeys. The time taken on the direct route from Royton to London is $6\frac{1}{2}$ hours. The time from Royton to Birmingham is $1\frac{1}{2}$ hours and Birmingham to London 2 hours. Hence we have a triangle with sides of length $1\frac{1}{2}$, 2 and $6\frac{1}{2}$! The right-hand diagram is an impossible triangle.

This is acceptable because the diagram shows a network representing a real-life situation and not a scale drawing.

> Real-life problems may not obey the triangle inequality.

2.3 Dijkstra's algorithm

In 1959, Edsger Dijkstra invented an algorithm for finding the shortest path through a network. The following is a simple set of instructions that enables students to follow the algorithm:

Step 1 Label the start vertex as 0.

Step 2 Box this number (permanent label).

Step 3 Label each vertex that is connected to the start vertex with its distance (temporary label).

Step 4 Box the smallest number.

Step 5 From this vertex, consider the distance to each connected vertex.

Step 6 If a distance is less than a distance already at this vertex, cross out this distance and write in the new distance. If there was no distance at the vertex, write down the new distance.

Step 7 Repeat from step 4 until the destination vertex is boxed.

Sometimes edges of networks have arrows that show you which directions you must go from one vertex to the next. These are called **directed networks**. To apply Dijkstra's algorithm to a directed network, at each stage only consider edges that lead **from** the vertex.

Note: When a vertex is boxed you do not reconsider it. You need to show all temporary labels together with their crossings out.

If a vertex is boxed then you do not write down a new temporary value. This would be a complete enumeration.

Worked example 2.1

Find the shortest distance from A to J on the network below.

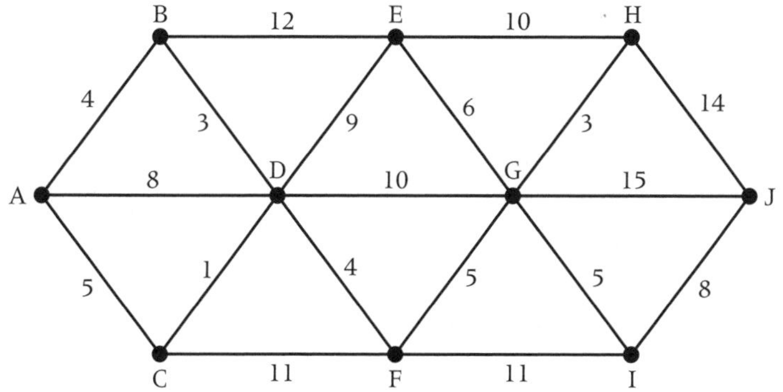

Solution

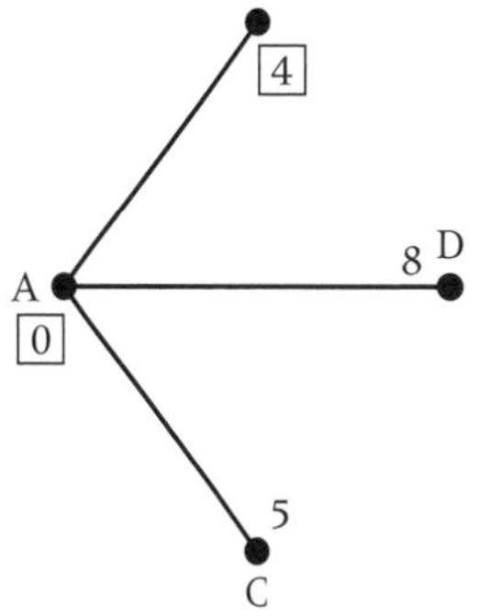

Step 1 Label A as 0.

Step 2 Box this number.

Step 3 Label values of 4 at B, 8 at D and 5 at C.

Step 4 Box the 4 at B.

Step 5 From B, the connected vertices are D and E. The distances at these vertices are 7 at D (4 + 3) and 16 at E (4 + 12).

Step 6 As the distance at D is 7, lower than the 8 currently at D, cross out the 8.

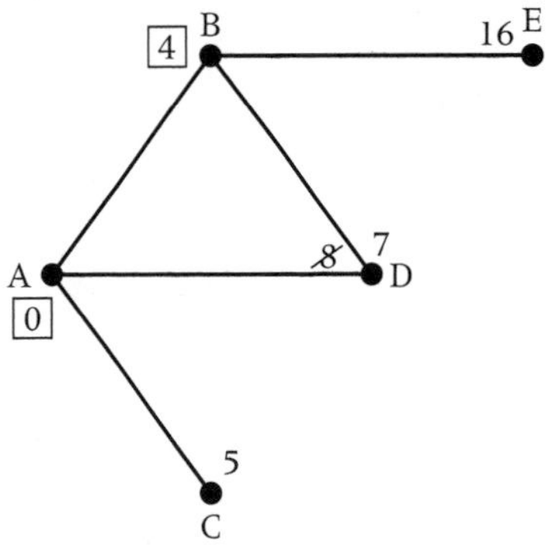

Step 4 Box the smallest number, which is the 5 at C.

Step 5 From C, the connected vertices are D and F. The distances at these vertices are 6 at D (5 + 1) and 16 at F (5 + 11).

Step 6 As the distance at D is 6, lower than the 7 currently at D, cross out the 7.

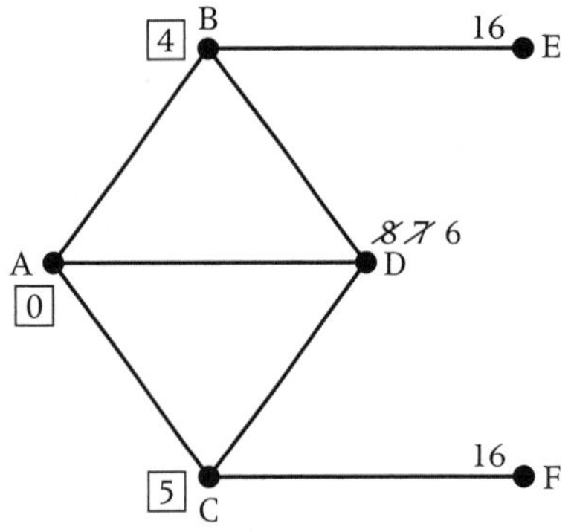

Step 4 Box the smallest number, which is the 6 at D.

Step 5 From D, the connected vertices are E, F and G. The distances at these vertices are 15 at E (6 + 9), 16 at G (6 + 10) and 10 at F (6 + 4).

Step 6 As the distance at E is 15, lower than the 16 currently at E, cross out the 16. As the distance at F is 10, lower than the 16 currently at F, cross out the 16.

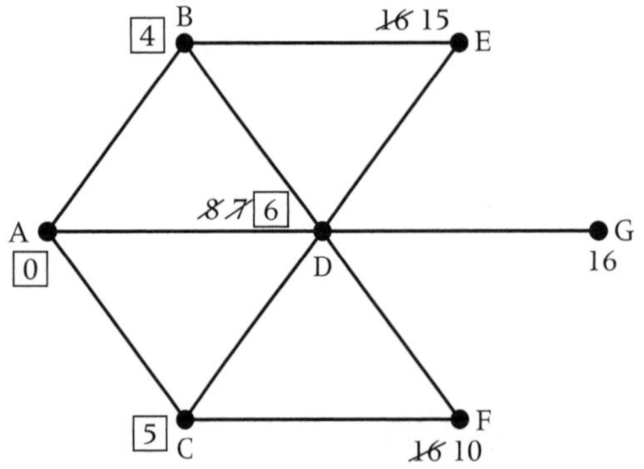

Step 4 Box the smallest number, which is the 10 at F.

Step 5 From F, the connected vertices are G and I. The distances at these vertices are 15 at G (10 + 5) and 21 at I (11 + 10).

Step 6 As the distance at G is 15, lower than the 16 currently at G, cross out the 16.

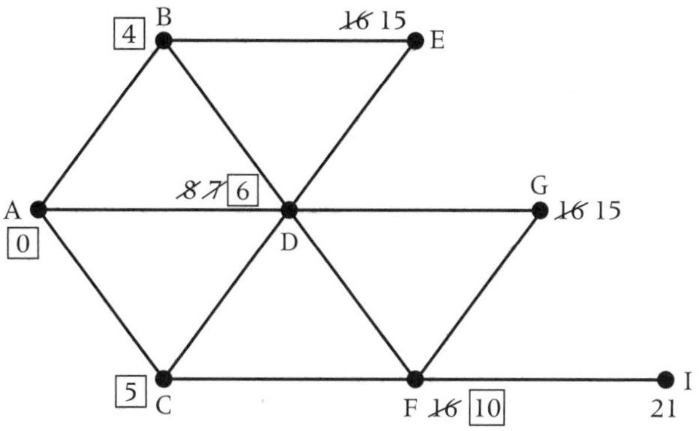

Step 4 Box the smallest number, which the 15 at either E or G (it doesn't matter which you chose).

Step 5 From E, the connected vertices are H and G. The distances at these vertices are 21 at G (15 + 6) and 25 at H (15 + 10). Do not write down the value of 21 at G as this is greater than the number already there.

Step 6 There are no improvements, so there is no crossing out.

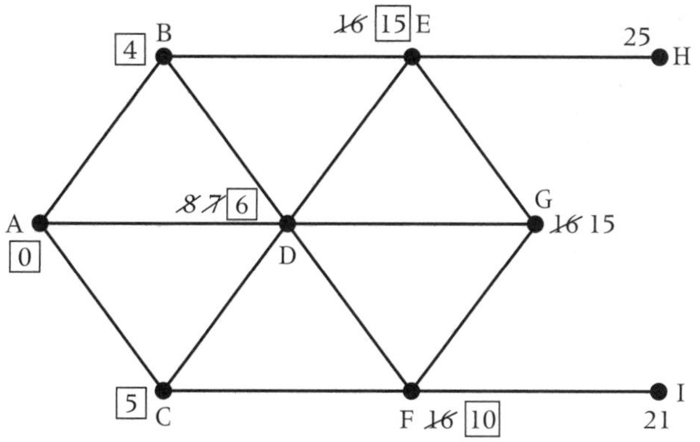

Step 4 Box the smallest number, which is the 15 at G.

Step 5 From G, the connected vertices are H, I and J. The distances at these vertices are 18 at H (15 + 3), 20 at I (15 + 5) and 30 at J (15 + 15).

Step 6 As the distance at H is 18, lower than the 25 currently at H, cross out the 25. As the distance at I is 20, lower than the 21 currently at I, cross out the 21.

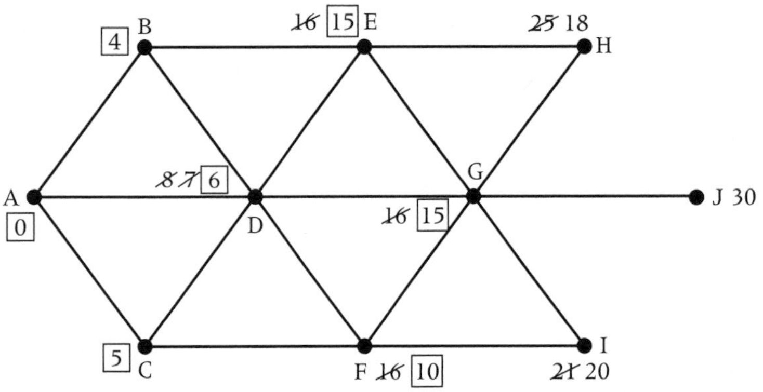

Step 4 Box the smallest number, which is the 18 at H.

Step 5 From H, the connected vertex is J. The distance at this vertex is 32 (18 + 14). Do not write down the value of 32 at J as this is greater than the 30 already there.

Step 6 There are no improvements, so there is no crossing out.

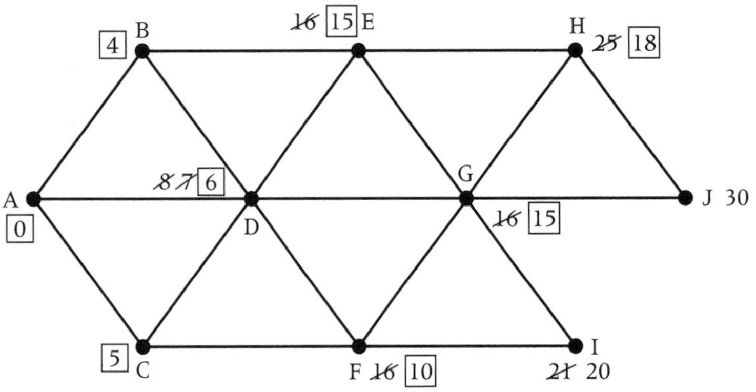

Step 4 Box the smallest number, which is the 20 at I.

Step 5 From I, the connected vertex is J. The distance at this vertex is 28 (20 + 8).

Step 6 As the distance at J is 28, lower than the 30 currently at J, cross out the 30.

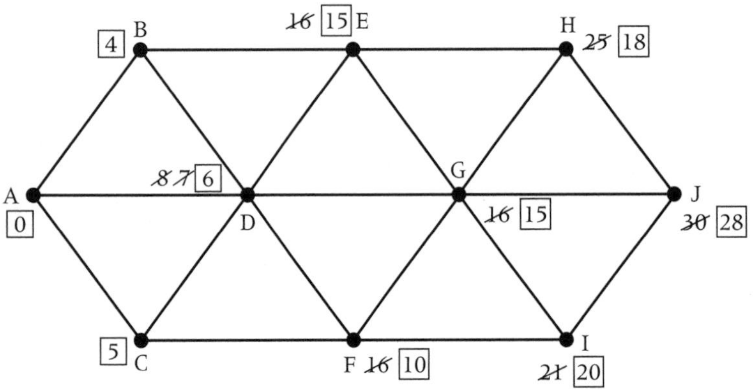

Step 7 The final vertex, in this case J, is not boxed. The boxed number at J is the shortest distance. The route corresponding to this distance of 28 is ACDFGIJ, but this is not immediately obvious from the network.

How do we retrace the route that corresponds to this shortest network? This problem will be dealt with later in the chapter.

Worked example 2.2

Find the shortest distance from A to H on the network below.

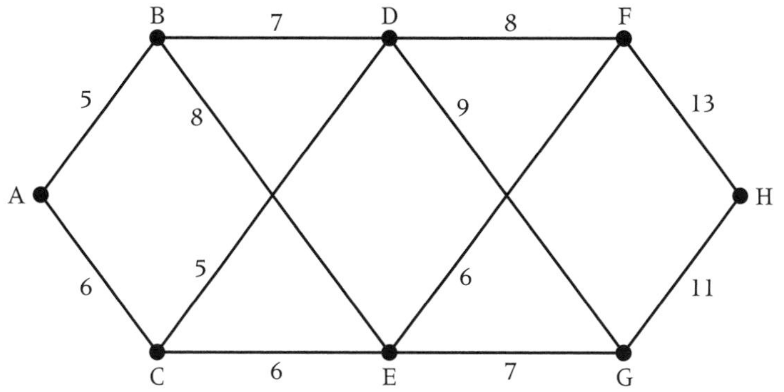

Solution

The fully labelled diagram below shows the values, both temporary and permanent, at each vertex.

Any answer to an exam question must have exactly the same amount of detail as shown here.

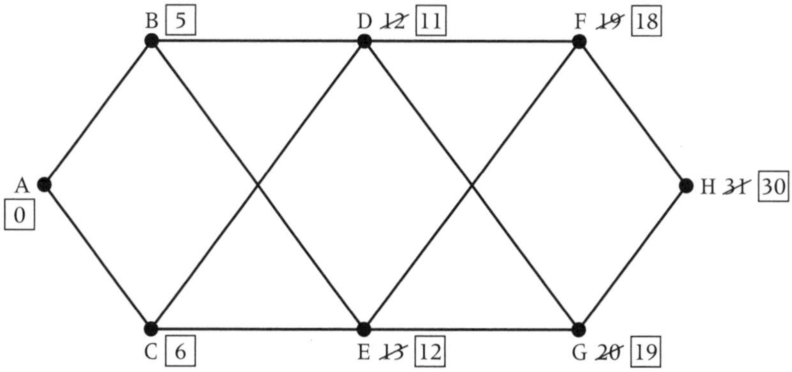

EXERCISE 2A

1 Use Dijkstra's algorithm on the networks below to find the shortest distance from A to H.

(a)

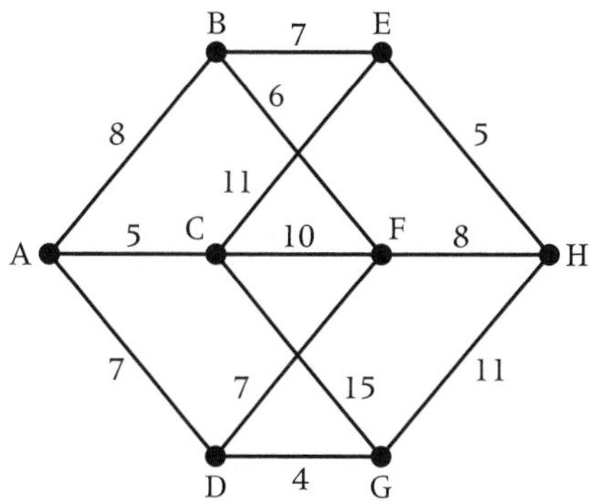

(b)

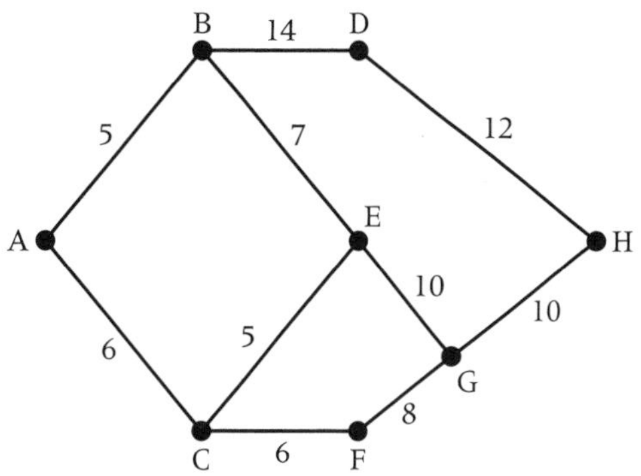

2 The diagram below shows roads connecting towns near to Rochdale. The numbers on each arc represent the time, in minutes, required to travel along each road. Peter is delivering books from his base at Rochdale to Stockport. Use Dijkstra's algorithm to find the minimum time for Peter's journey.

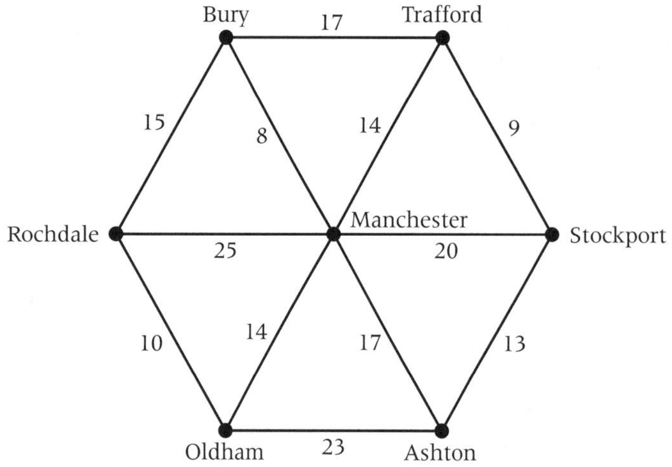

3 The diagram below shows roads connecting villages near to Royton. The numbers on each arc represent the distance, in miles, along each road. Leon lives in Royton and works in Ashton. Use Dijkstra's algorithm to find the minimum distance for Leon's journey to work.

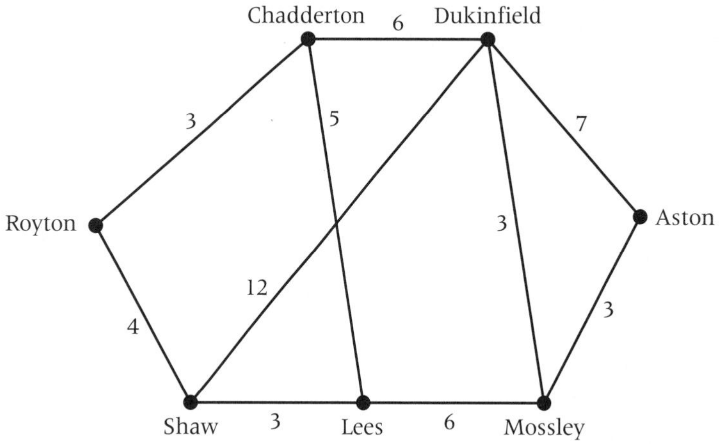

2.4 Multiple start points

How do we cope with a situation in which instead of having a single starting point and a single end point there are multiple start points and a single end point? For example, in the Monte Carlo rally cars start from a number of different countries but all end up at the same finishing point. Dijkstra's algorithm gives a method of finding the shortest distance, as in the previous example from A to J, but it is an identical problem to find the shortest distance from J to A.

If there are multiple start points, then we apply Dijkstra's algorithm from the end point until we have reached each of the starting points. In this way we can find the shortest route.

Worked example 2.3

Three boys walk to school, J, from their homes at A, B and C. The diagram shows the network of roads near their homes and school.
The numbers on each arc represent the distance, in metres, along each road. Use Dijkstra's algorithm from J to find which boy lives nearest to the school.

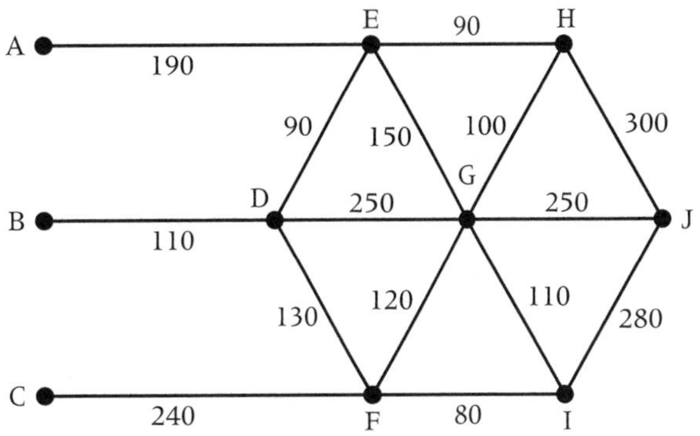

Solution

If we were to apply Dijkstra's algorithm in the normal way, the workings would be difficult to follow. There would be temporary and permanent labels from each of the three starting points. Working backwards from J, and applying the standard algorithm, the worked solution below is obtained.

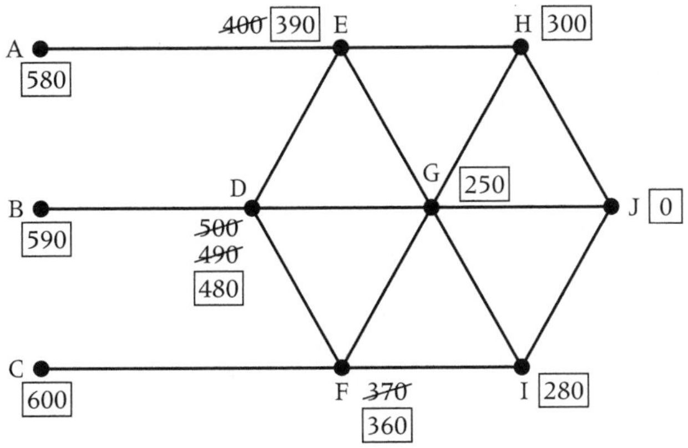

The boy living in house A lives closest to the school.

EXERCISE 2B

1 Use Dijkstra's algorithm, starting from H, on the networks below to find which of the vertices A or B is nearer to H.

(a)

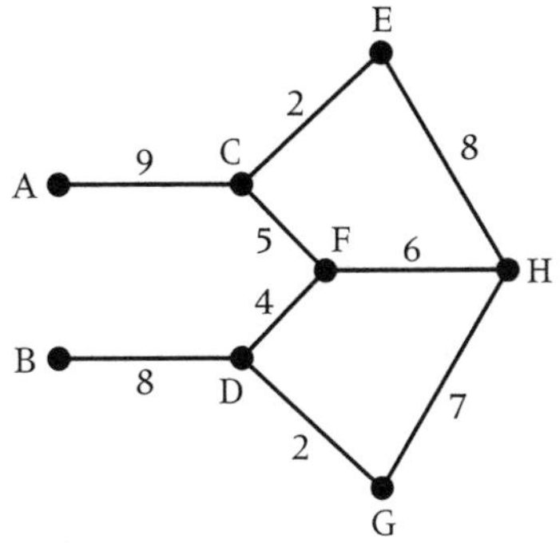

(b)

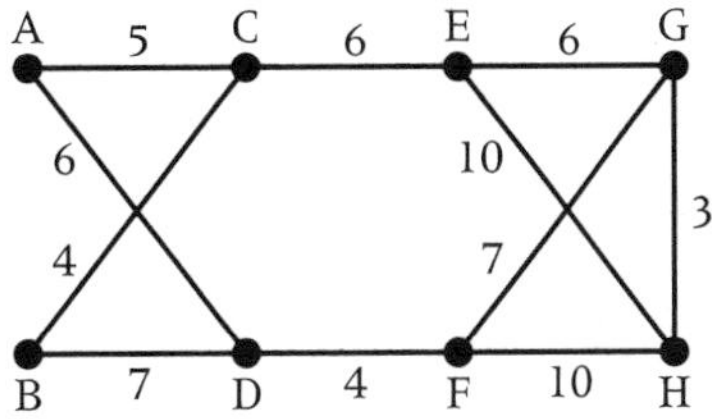

2.5 Limitations

Whilst Dijkstra's algorithm gives an exact method of finding the shortest distance connecting two points, it doesn't work if any of the edges have a negative value. How can you have an edge of negative length? If we are talking about distances then obviously something like this is impossible, but consider the following situation.

A catalogue delivery firm calculates the cost of delivering along a variety of routes. Each of the distances is represented by an actual cost. However, if whilst on a delivery a delivery van also makes a collection, then the value of making this collection en route may save more money than it would normally cost by driving along this road. Hence you may have an edge that has a negative value.

> If we use Dijkstra's algorithm on a network containing an edge that has a negative value it does not work.

If we accept that negative lengths are possible, then it will become immediately obvious that Dijkstra's algorithm won't work because using the algorithm we don't revisit anywhere that has already been boxed, yet if we come to a vertex then from there a negative length may produce a value that is shorter than one that is already boxed. In such situations, where Dijkstra's algorithm fails, the normal approach is to use dynamic programming (see D2).

Worked example 2.4

The following network shows the cost, in pounds, of travelling along a series of roads. Use Dijkstra's algorithm to find the minimum cost of travelling from A to F.

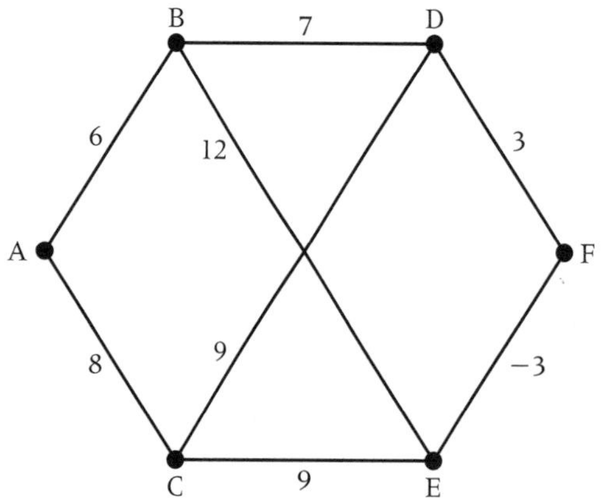

Solution

Working through the network in the standard way, the worked diagram below is obtained.

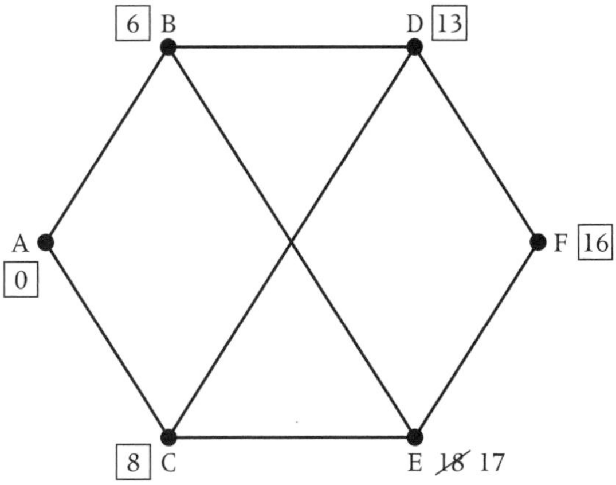

At F, the figure of 16 is boxed and hence the question is finished. However, from E to F a value of -3 reduces the minimum cost of getting to F to 14!

2.6 Finding the route

We can use Dijkstra's algorithm to find the length of the shortest route between two vertices in a network. However, if we want to find the corresponding route, we need to record more information as we apply the algorithm.

> Instead of listing temporary values, we put a letter after each value, which indicates the preceding vertex on the route. We find the route by backtracking through the network from the finishing point.

Worked example 2.5

The network below was used in Worked example 1.

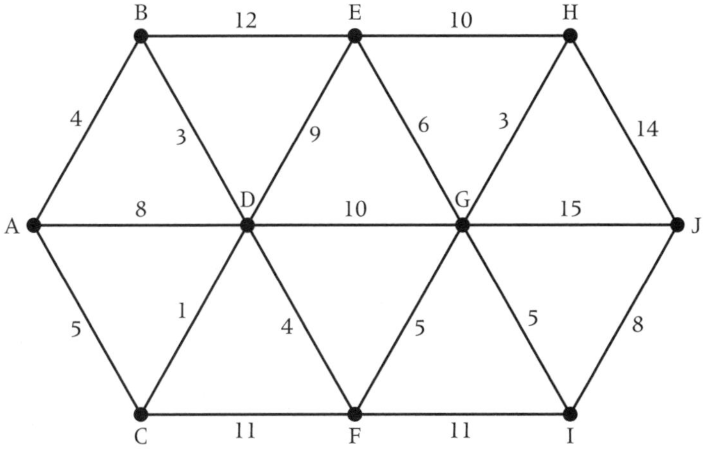

Solution

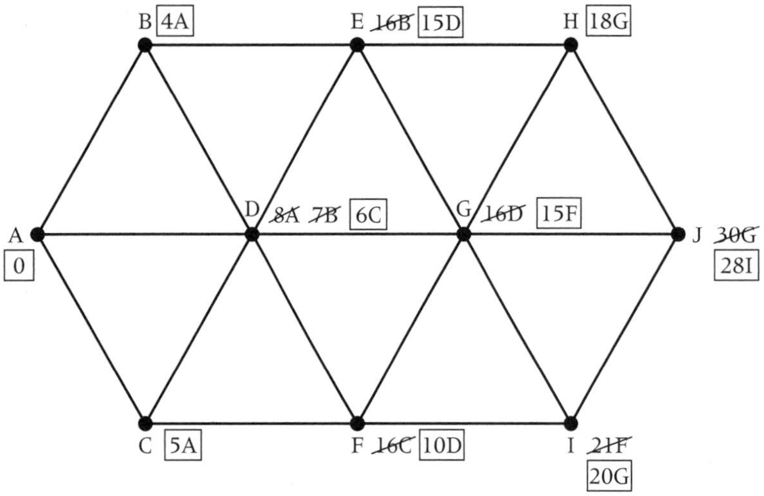

Box A has a value of zero.
At B, the distance from A is 4 so we now write 4A.
At D, the distance from A is 8 so we write 8A and at C we write 5A.
We then box the value of 4 at B, so 4A is now boxed.
From B, the value at E becomes 16B. At D we get 7B.
Box the smallest number, which is the value of 5A at C, so 5A is boxed at C.
From there, there is a value of 16C at F and a value of 6C at D.
We box the smallest value, which is 6C at D.
At D there is 8A crossed out, 7B crossed out and 6C, which has been boxed. This tells us that to get to D the smallest distance is 6 and we came from vertex C.
Working from the finishing point J we have a boxed value of 28I so we now look at vertex I. Here the boxed value is 20G so we now look at vertex G, and so on until we return to A. Hence the shortest path is ACDFGIJ, with length 28.

EXERCISE 2C

Repeat Exercises 2A and 2B to find the routes that correspond to the minimum distances.

MIXED EXERCISE

1

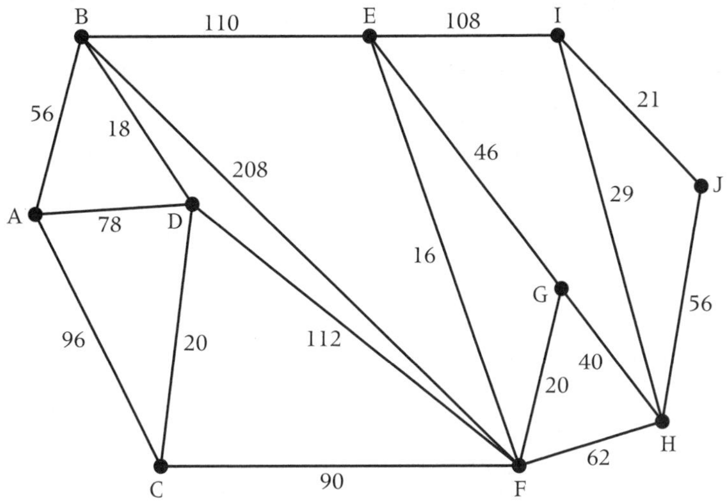

(a) Use Dijkstra's algorithm on the above diagram to find the minimum time to travel from A to J, and state the route.

(b) A new road is to be constructed connecting B to G. Find the time needed for travelling this section of road if the overall minimum journey time to travel from A to J is reduced by 10 minutes. State the new route. [A]

2 The following network shows the time, in minutes, of train journeys between seven stations.

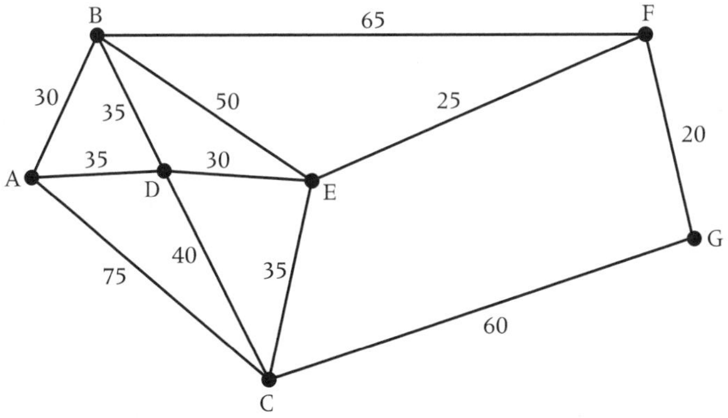

(a) Given that there is no time delay in passing through a station, use Dijkstra's algorithm to find the shortest time to travel from A to G.

(b) Find the shortest time to travel from A to G if in reality each time the train passes through a station, excluding A and G, an extra 10 minutes is added to the journey time. [A]

3 The following network shows two islands, each with 7 small towns. One road bridge connects the two islands. Values shown represent distances by road, in miles.

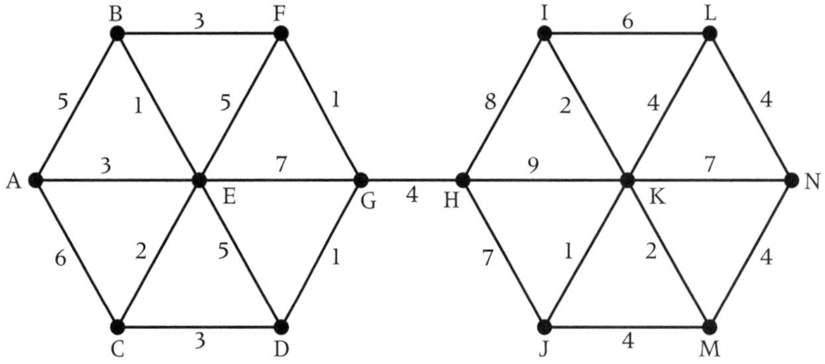

Use Dijkstra's algorithm to find the shortest distance between A and N, stating the route. [A]

4 The following diagram shows main roads connecting places near to Manchester, where the values shown represent the distances in miles. Mark lives in Rochdale and works in Trafford.

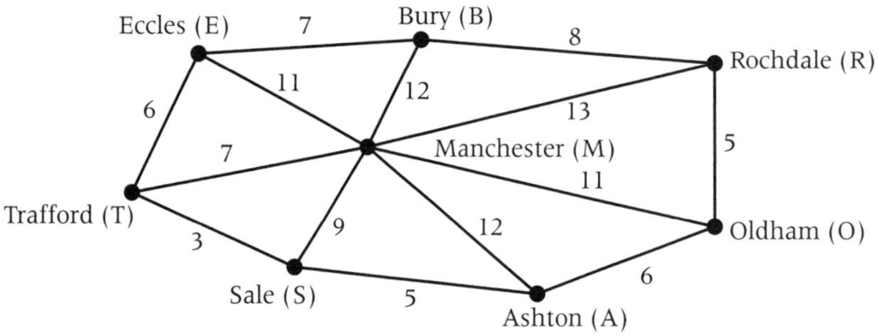

(a) Use Dijkstra's algorithm to find the shortest distance from Rochdale to Trafford. Write down the corresponding route.

(b) A new orbital motorway is built around Manchester as shown in the diagram below. The values shown represent the distances in miles.

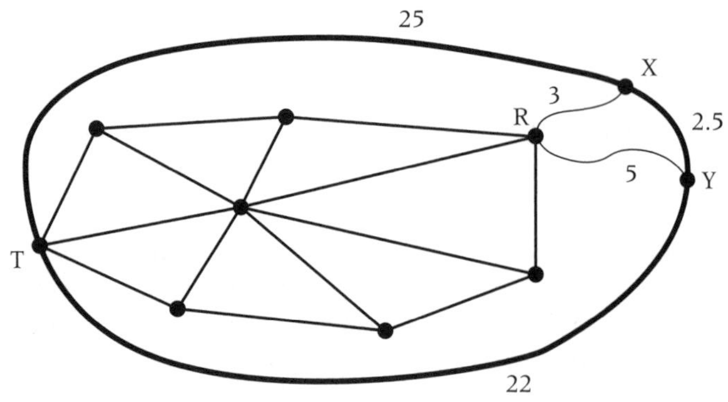

Mark has access to the new motorway at points X and Y. Due to traffic conditions, he can drive at 20 mph on all main roads and 40 mph on the motorway.
Find the minimum time for Mark to travel from Rochdale to Trafford and state the route he should take. [A]

5 Every day, Mary thinks of a rumour to spread on her way to school. The rumour is then spread from one person to another. The following network shows the route through which the rumour spreads. The number on each arc represents the time, in minutes, for the rumour to spread from one person to another.

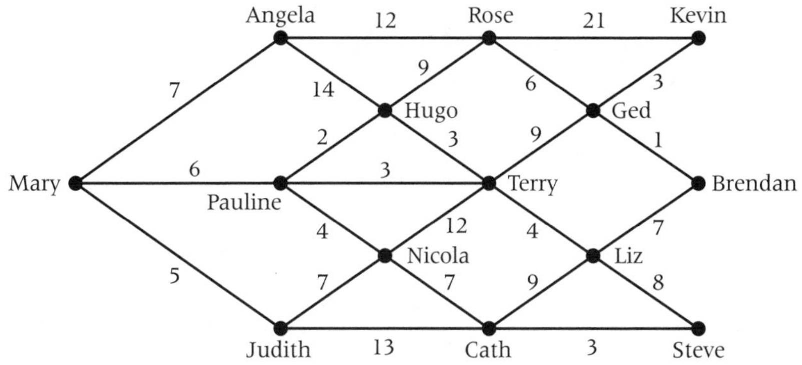

(a) Use Dijkstra's algorithm to find the time taken for the rumour to reach each person.

(b) List the route through which Brendan first hears the rumour.

(c) On a particular day Pauline is not at school. Find, by inspection, the extra time that elapses before Brendan first hears the rumour for that day. [A]

6 Three boys, John, Lee and Safraz, are to take part in a running race. They are each starting from a different point but they all must finish at the same point N.

John starts from the point A, Lee from the point B and Safraz from the point C.

The following diagram shows the network of streets that they may run along. The numbers on the arcs represent the time, in seconds, taken to run along each street.

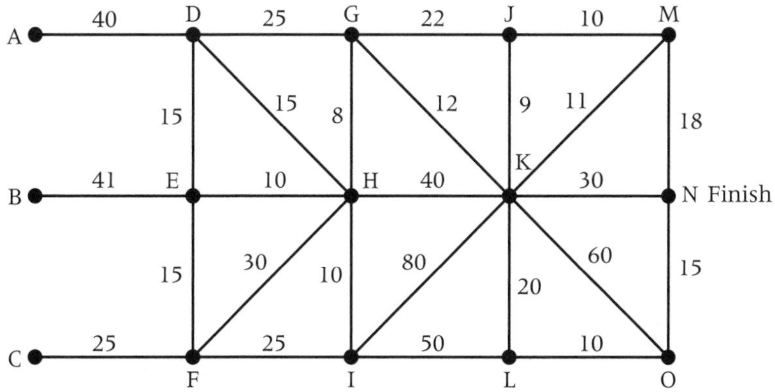

(a) Working backwards from N, or otherwise, use Dijkstra's algorithm to find the time taken for each of the three boys to complete the course. Show all your working at each vertex.

(b) Write down the route that each boy should take. [A]

7 A school is organising a short road race in which pupils have to start at A and use their own choice of route to reach J as quickly as possible. The diagram below shows the network of roads available and, for each road, the minimum completion time, in seconds, for a girl in the school.

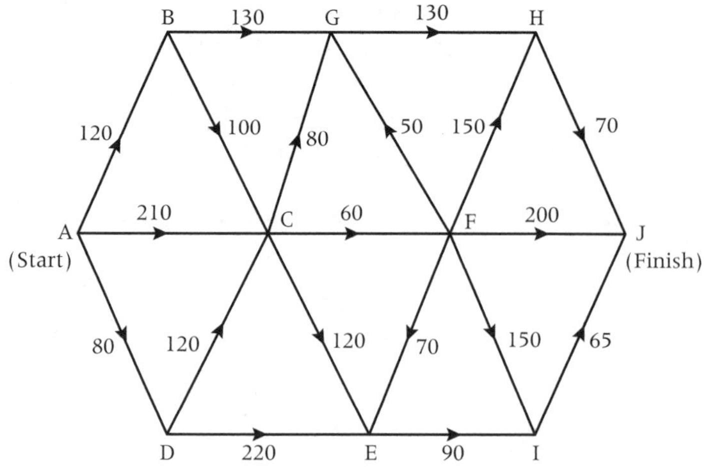

Mia, the best girl runner in the school, is planning her strategy. Given that she is able to run each section in the minimum time, use Dijkstra's algorithm to find the route she should take and her time for the race. [A]

Hint for Question 7: See margin note on page 25.

8 An insurance salesman has to drive from his home at A to his head office at L. The time, in minutes, for each section of the journey is shown in the diagram below.

(a) Use Dijkstra's algorithm to find the minimum time for the total journey and state the route the salesman should take.

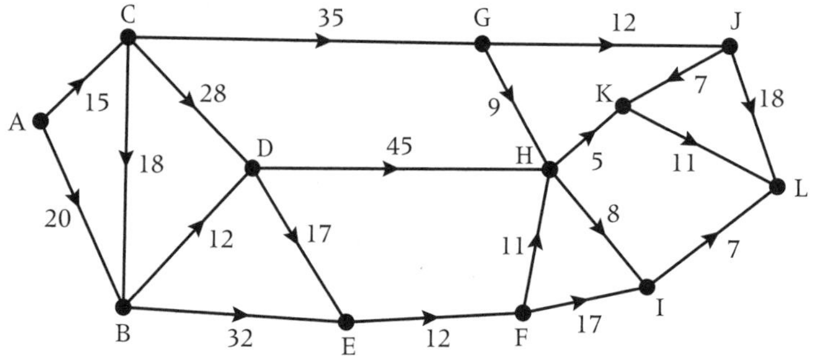

(b) A new road is constructed joining D to F, as shown below. The journey time for this section of road is x minutes. Find an expression, in terms of x, for the minimum time for the journey from A to L using the new road. [A]

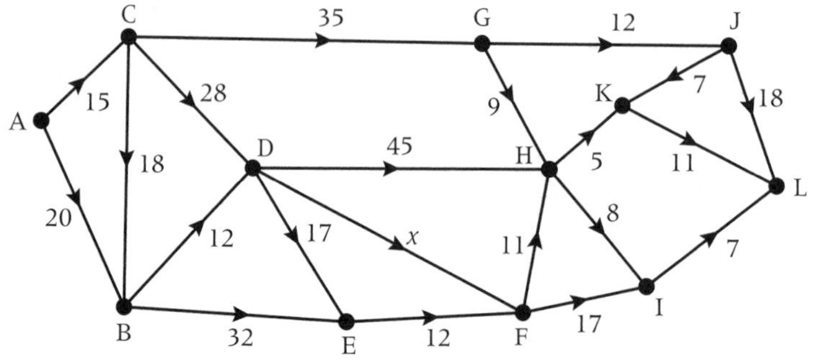

9 The following diagram shows motorways and main roads connecting towns in Sicily. The numbers represent the times taken, in minutes, to drive along each road. There are two airports on the island, one at Catania (C) and one at Palermo (P).

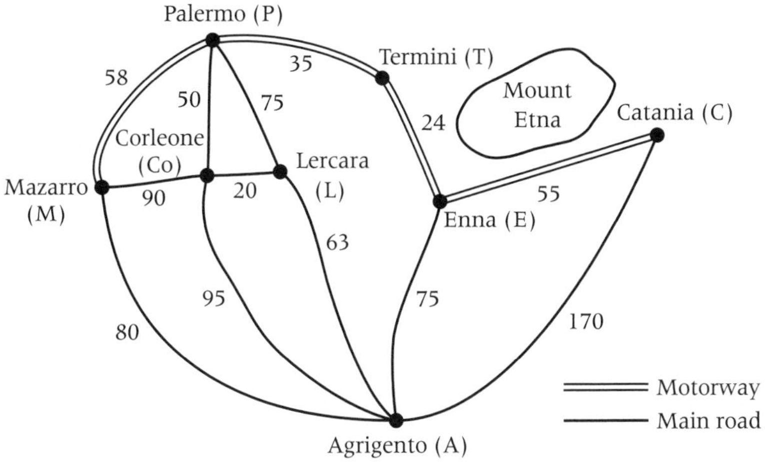

Stella plans to fly to Catania and then drive to Agrigento.

(a) Find, by inspection, the minimum time for Stella to drive from Catania to Agrigento.

(b) Due to the volcano at Mount Etna erupting, Stella's flight is diverted to Palermo.

  (i) Use Dijkstra's algorithm to find the minimum time to drive from Palermo to Agrigento.

  (ii) State the route that she should take.

(c) Stella drives at 50 km/h on main roads and 100 km/h on motorways. Given that she keeps her driving time to a minimum, find the extra distance that she would have had to drive if she had landed at Catania airport rather than at Palermo airport. [A]

10 The following diagram shows the lengths, in miles, of roads connecting 10 towns.

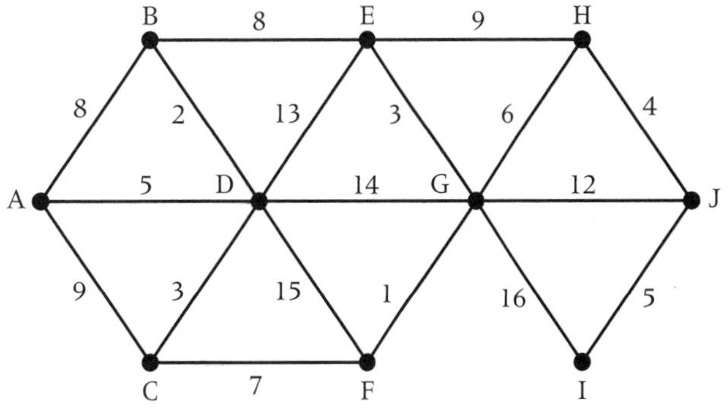

(a) Use Kruskal's algorithm, showing the order in which you select the edges, to find the minimum spanning tree for the network. Draw your minimum spanning tree and state its length.

(b) **(i)** Use Dijkstra's algorithm to find the shortest distance from A to J. State the route corresponding to this minimum distance.

  (ii) A new road is built connecting F to I. The length of this road is x miles, where x is an integer. A shorter route from A to J than that found in **(b)(i)** is now available.
  Find the maximum value of x. [A]

11 A railway company is considering opening some new lines between seven towns A–G. The possible lines and the cost of setting them up (in millions of pounds) are shown in the following network.

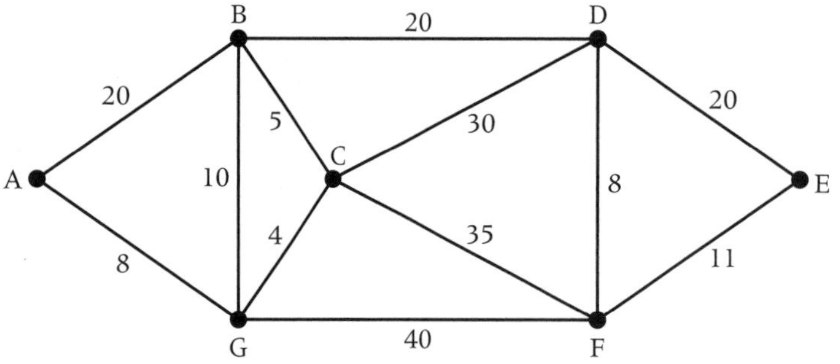

(a) Use Dijkstra's algorithm to find the minimum cost of opening lines from A to E. Show all your workings at each vertex.

(b) From your workings in **(a)** write down the minimum cost of opening lines:

(i) from A to B

(ii) from A to F.

(c) Use Kruskal's algorithm to find the minimum cost of opening lines so that it is possible to travel between any two of the towns by rail, and state the lines which should be opened in order to achieve this minimum cost.

(d) The rail company wants to open some of the lines so that it is possible to travel by rail starting at one town, finishing at another and passing through each of the other five towns exactly once. Which lines should the rail company open in order to do this as cheaply as possible? [A]

12 The diagram represents the roads joining 10 villages, labelled A to J. The numbers give distances in kilometres.

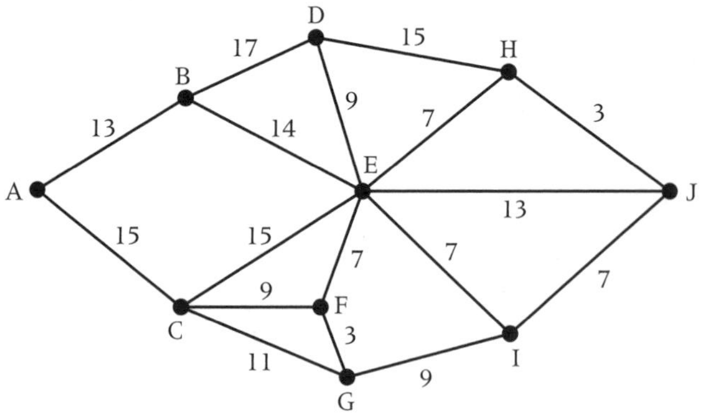

(a) Use Dijkstra's algorithm to find a shortest route from A to J. Explain the method carefully, and show all your working. Give a shortest route and its length.

(b) A driver usually completes this journey driving at an average speed of 60 km/h. The local radio reports a serious accident at village E, and warns drivers of a delay of 10 minutes.

Describe how to modify your approach to **(a)** to find the quickest route, explaining how to take account of this information. What is the quickest route, and how long will it take? [A]

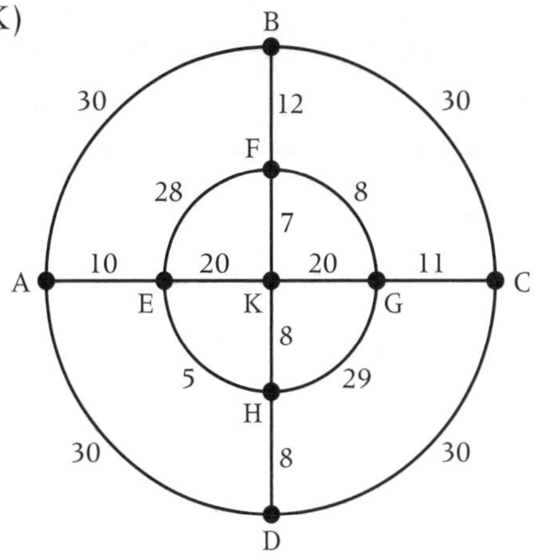

13 The network shows the roads around the town of Kester (K) and the times, in minutes, needed to travel by car along those roads.

(a) A motorist wishes to travel from A to C along these roads in the minimum possible time. Use Dijkstra's algorithm to find the route the motorist should use and the time that the journey will take. Show all your workings clearly.

(b) The four sections of ring road AB, BC, CD and DA each require the same amount of time, and next year there will be improvements to the ring road in order to reduce this time from 30 minutes to *m* minutes. This will enable the motorist to reduce the minimum time for a journey from A to C by 2 minutes. Find the value of *m* and state his new route. [A]

14 The network shows the distances in kilometres of various routes between points S, T, U, V, W, X, Y and Z.

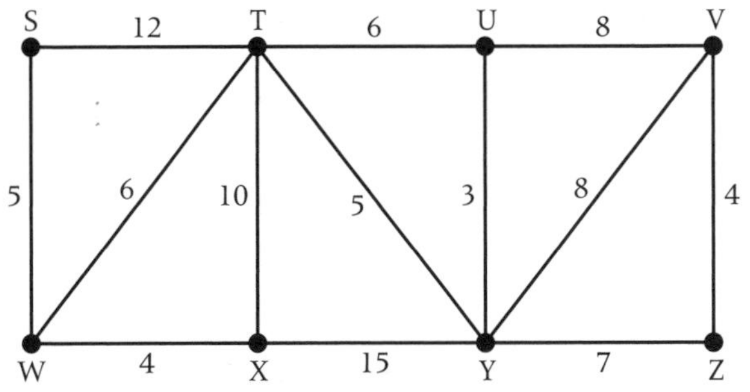

Use Dijkstra's algorithm to find the shortest path from S to Z. Show your working. [A]

Key point summary

1 Real-life problems may not obey the triangle inequality. *p24*

2 **Dijkstra's algorithm** enables the shortest path between two points to be found. *p25*

3 Dijkstra's algorithm is equally valid when used backwards through a network. *p30*

4 Dijkstra's algorithm fails if there are negative edges in a network. *p32*

5 Tracing a route through a network can be easily found if careful labelling of Dijkstra's algorithm is used. *p33*

Test yourself	What to review
1 Use Dijkstra's algorithm to find the shortest distance from A to E in the networks below.	*Section 2.3*

(a)

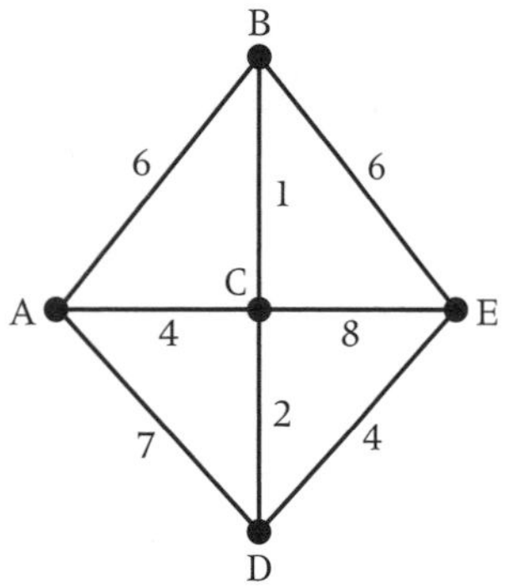

(b)

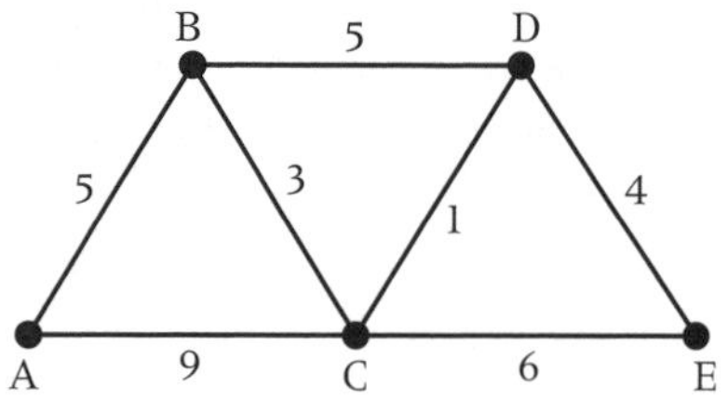

2 Find the route corresponding to the shortest distance for question 1.	*Section 2.6*
3 The network below has four vertices and five edges.	*Sections 2.3, 2.6*

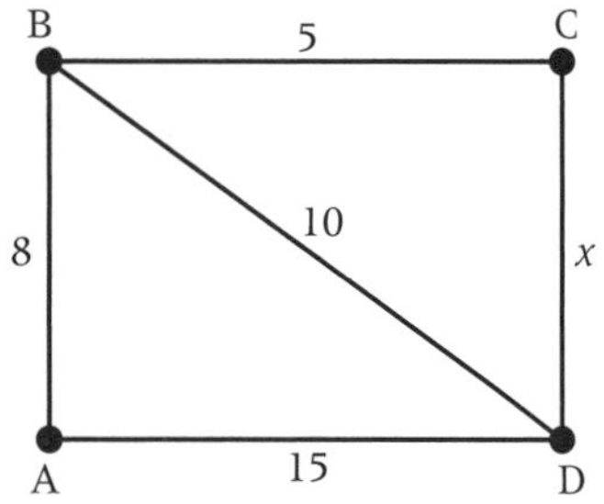

Given that the second shortest route from A to D is ABCD, find the range of values of x.

4 The network below has eight vertices and nine edges.	*Section 2.4*

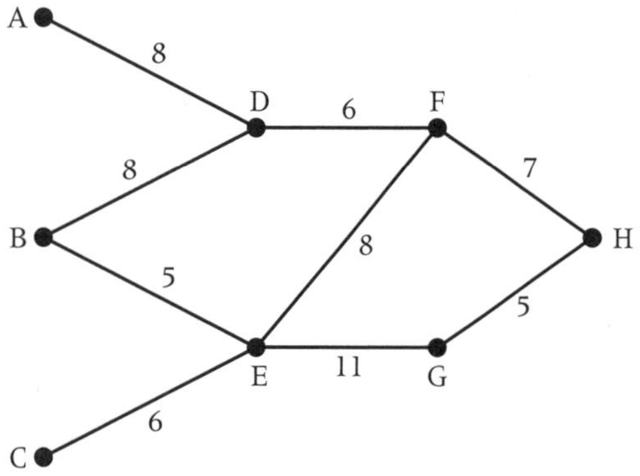

Find which of the vertices A, B or C is nearest to vertex H.

Test yourself ANSWERS

1 **(a)**

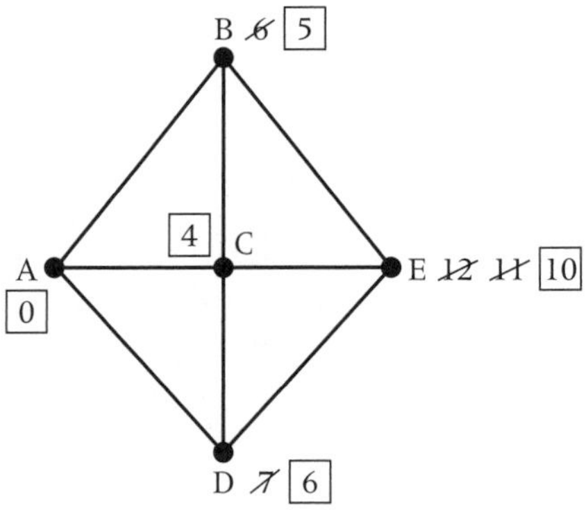

(b)

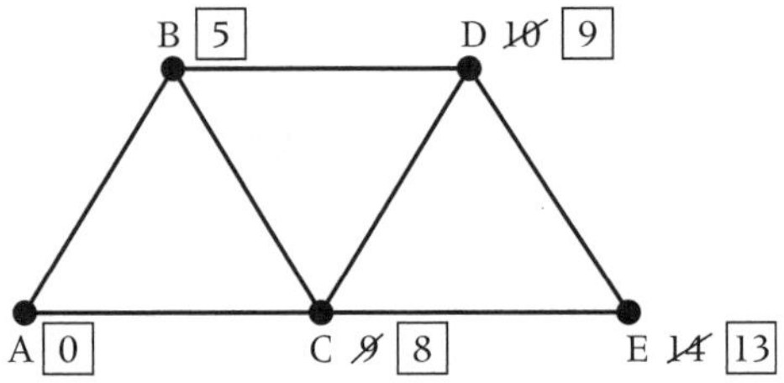

2 **(a)** ACDE

(b) ABCDE

3 $2 < x < 5$

4

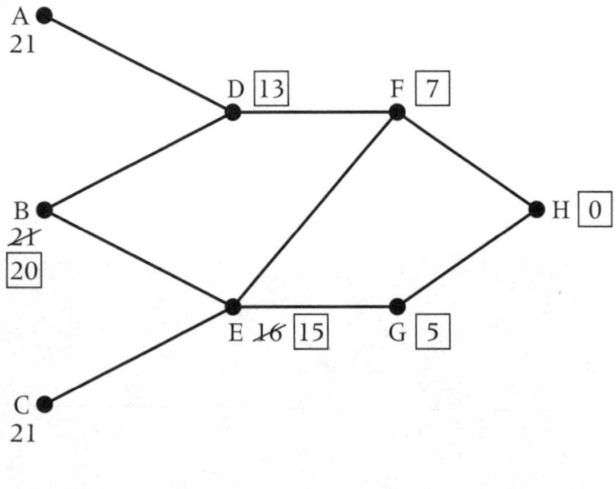

CHAPTER 3

Chinese postman problem

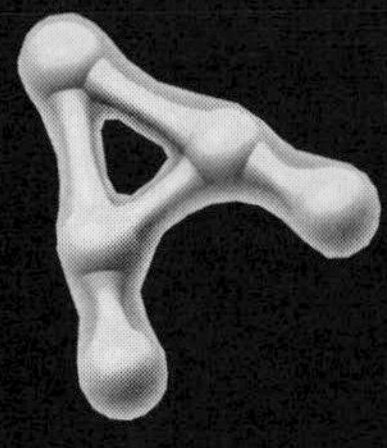

Learning objectives

After studying this chapter, you should be able to:

- understand the Chinese postman problem
- apply an algorithm to solve the problem
- understand the importance of the order of vertices of graphs.

3.1 Introduction

In 1962, a Chinese mathematician called Kuan Mei-Ko was interested in a postman delivering mail to a number of streets such that the total distance walked by the postman was as short as possible. How could the postman ensure that the distance walked was a minimum?

In the following example a postman has to start at A, walk along all 13 streets and return to A. The numbers on each edge represent the length, in metres, of each street. The problem is to find a trail that uses all the edges of a graph with minimum length.

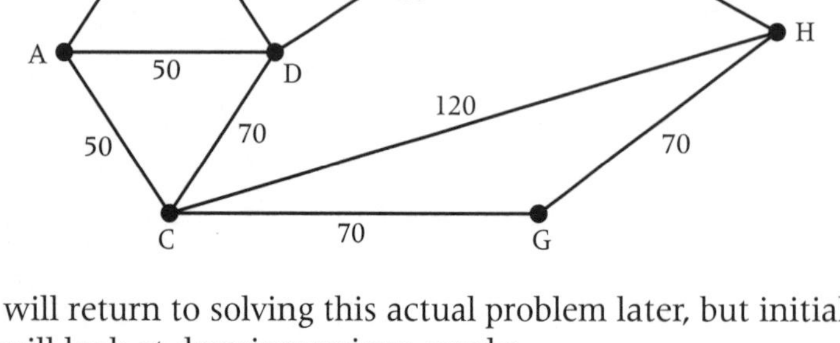

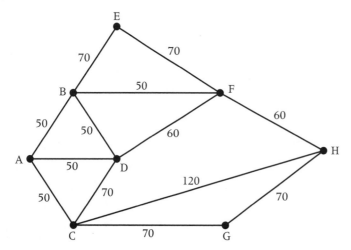

We will return to solving this actual problem later, but initially we will look at drawing various graphs.

3.2 Traversable graphs

> A **traversable** graph is one that can be drawn without taking a pen from the paper and without retracing the same edge. In such a case the graph is said to have an Eulerian trail.

Eulerian trails are dealt with in detail in Chapter 5.

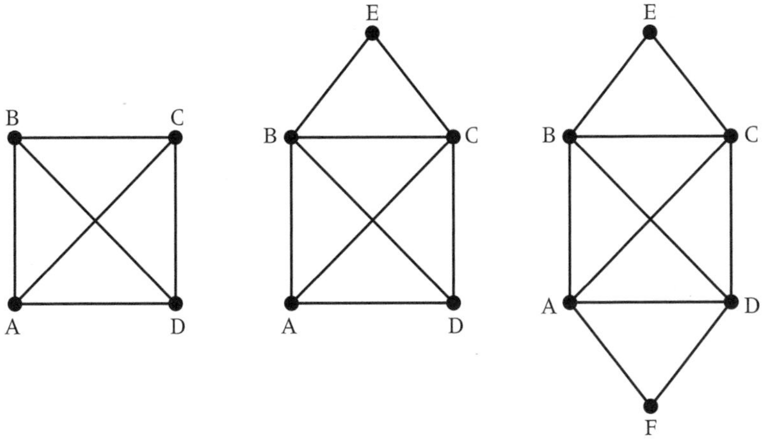

Graph 1 Graph 2 Graph 3

If we try drawing the three graphs shown above we find:

- it is impossible to draw Graph 1 without either taking the pen off the paper or re-tracing an edge
- we can draw Graph 2, but only by starting at either A or D – in each case the path will end at the other vertex of D or A
- Graph 3 can be drawn regardless of the starting position and you will always return to the start vertex.

What is the difference between the three graphs?
In order to establish the differences, we must consider the order of the vertices for each graph. We obtain the following:

Graph 1

Vertex	**Order**
A	3
B	3
C	3
D	3

Graph 2

Vertex	**Order**
A	3
B	4
C	4
D	3
E	2

Graph 3

Vertex	Order
A	4
B	4
C	4
D	4
E	2
F	2

When the order of all the vertices is even, the graph is traversable and we can draw it. When there are two odd vertices we can draw the graph but the start and end vertices are different. When there are four odd vertices the graph can't be drawn without repeating an edge.

> An **Eulerian** trail uses all the edges of a graph. For a graph to be Eulerian all the vertices must be of even order.
>
> If a graph has two odd vertices then the graph is said to be **semi-Eulerian**. A trail can be drawn starting at one of the odd vertices and finishing at the other odd vertex.

To draw the graph with odd vertices, edges need to be repeated. To find such a trail we have to make the order of each vertex even. In graph 1 there are four vertices of odd order and we need to pair the vertices together by adding an extra edge to make the order of each vertex four. We can join AB and CD, or AC and BD, or AD and BC.

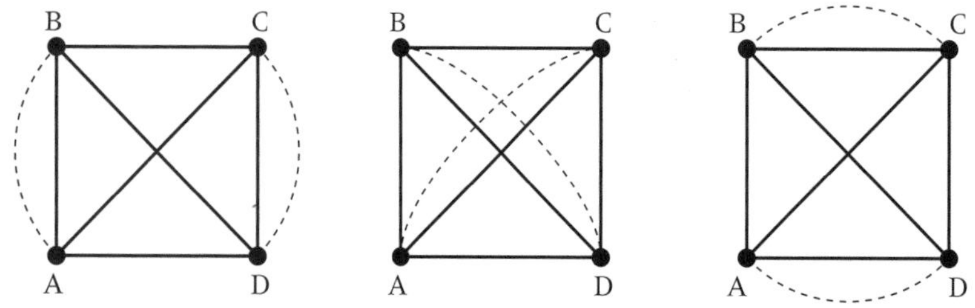

In each case the graph is now traversable.

Worked example 3.1

Which of the graphs below is traversable?

(a)

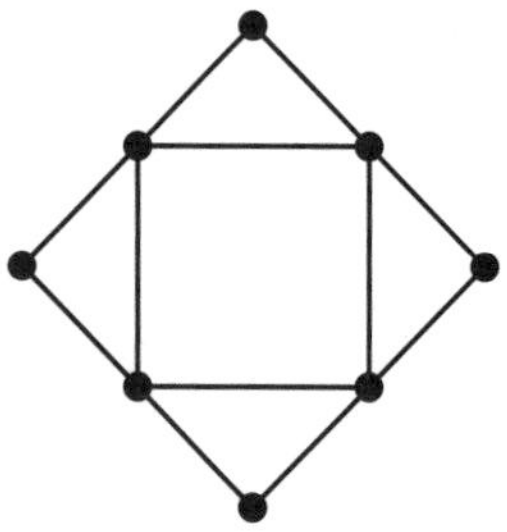

(b)

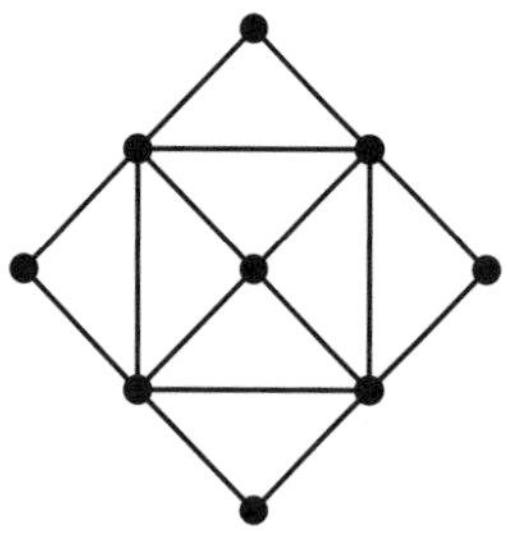

(c)

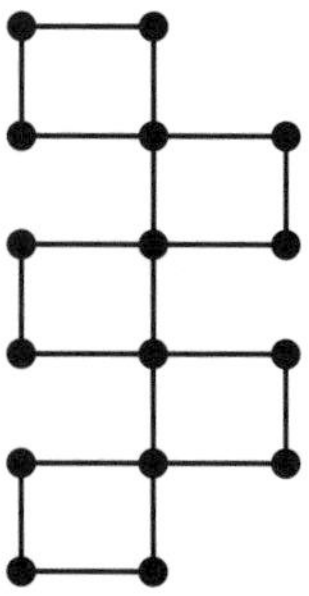

Solution

Graphs **(a)** and **(c)** are traversable as all the vertices are of even order. Graph **(b)** is not traversable as there are more than 2 vertices of odd order.

EXERCISE 3A

Which of the graphs below are traversable?

1

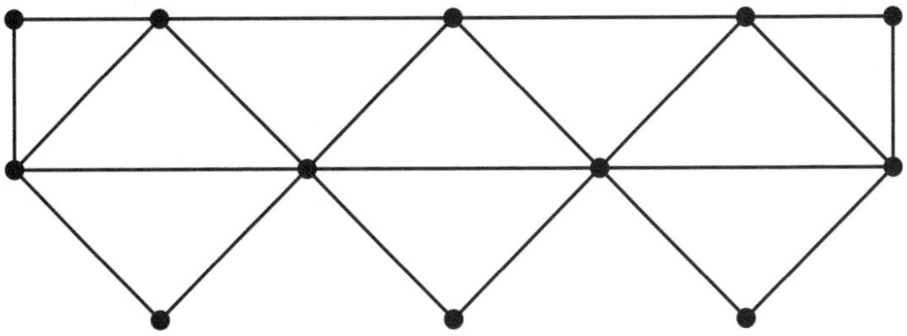

2

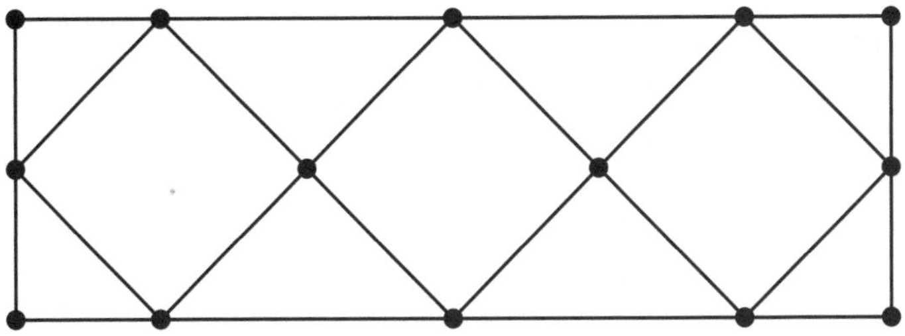

3

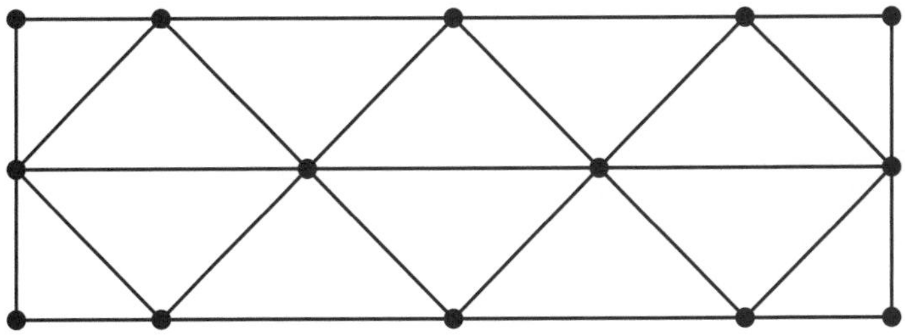

3.3 Pairing odd vertices

If there are two odd vertices there is only one way of pairing them together.
If there are four odd vertices there are three ways of pairing them together.
How many ways are there of pairing six or more odd vertices together?
If there are six odd vertices ABCDEF, then consider the vertex A. It can be paired with any of the other five vertices and still leave four odd vertices. We know that the four odd vertices can be paired in three ways; therefore the number of ways of pairing six odd vertices is $5 \times 3 \times 1 = 15$.

Similarly, if there are eight odd vertices ABCDEFGH, then consider the first odd vertex A. This could be paired with any of the remaining seven vertices and still leave six odd vertices. We know that the six odd vertices can be paired in 15 ways therefore the number of ways of pairing eight odd vertices is
$7 \times 5 \times 3 \times 1 = 105$ ways.
We can continue the process in the same way and the results are summarised in the following table.

Number of odd vertices	Number of possible pairings
2	1
4	$3 \times 1 = 3$
6	$5 \times 3 \times 1 = 15$
8	$7 \times 5 \times 3 \times 1 = 105$
10	$9 \times 7 \times 5 \times 3 \times 1 = 945$
n	$(n-1) \times (n-3) \times (n-5) \ldots \times 1$

Exam questions will not be set where candidates will have to pair more than four odd vertices but students do need to be aware of the number of ways of pairing more than four odd vertices.

3.4 Chinese postman algorithm

To find a minimum Chinese postman route we must walk along each edge at least once and in addition we must also walk along the least pairings of odd vertices on one extra occasion.

An algorithm for finding an optimal Chinese postman route is:

Step 1 List all odd vertices.

Step 2 List all possible pairings of odd vertices.

Step 3 For each pairing find the edges that connect the vertices with the minimum weight.

Step 4 Find the pairings such that the sum of the weights is minimised.

Step 5 On the original graph add the edges that have been found in Step 4.

Step 6 The length of an optimal Chinese postman route is the sum of all the edges added to the total found in Step 4.

Step 7 A route corresponding to this minimum weight can then be easily found.

Worked example 3.2

If we now apply the algorithm to the original problem:

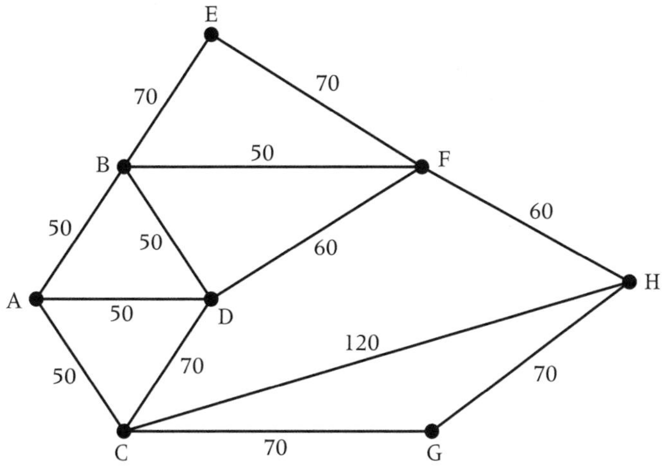

Step 1 The odd vertices are A and H.

Step 2 There is only one way of pairing these odd vertices, namely AH.

Step 3 The shortest way of joining A to H is using the path AB, BF, FH, a total length of 160.

Step 4 Draw these edges onto the original network.

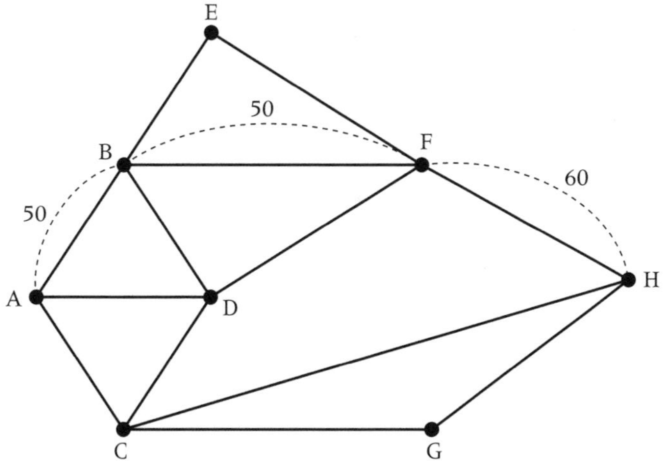

Step 5 The length of the optimal Chinese postman route is the sum of all the edges in the original network, which is 840 m, plus the answer found in Step 4, which is 160 m. Hence the length of the optimal Chinese postman route is 1000 m.

Step 6 One possible route corresponding to this length is ADCGHCABDFBEFHFBA, but many other possible routes of the same minimum length can be found.

EXERCISE 3B

1 Find the length of an optimal Chinese postman route for the networks below.

(a)

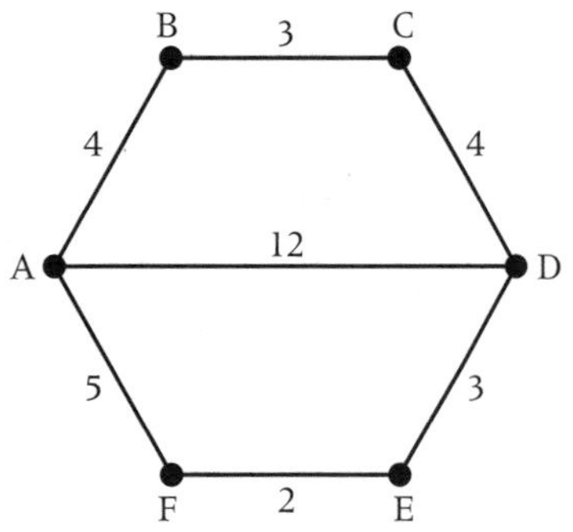

(b)

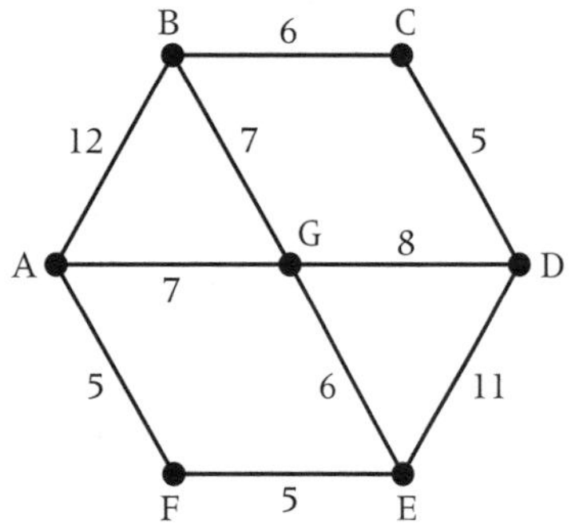

(c)

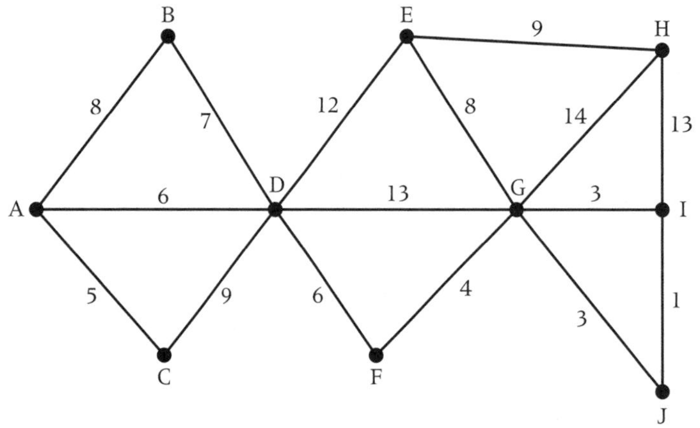

3.5 Finding a route

The method for finding the length of the Chinese postman route is quite straightforward, but to find the list of edges corresponding to this route can be quite tricky, especially in complicated networks. It is useful to calculate how many times each vertex will appear in a Chinese postman route. The following method should be applied before trying to find the route.

Step 1 On the original diagram add the extra edges to make the graph Eulerian.

Step 2 List the order of each vertex. At this stage each vertex will have an even order.

Step 3 The number of times each edge will appear in a Chinese postman route will be half the order of its vertex, with the exception being vertex A (the start/finish vertex), as this will appear on one extra occasion.

Referring to the diagram below, the orders of the vertices are as follows:

Vertex	Order
A	4
B	6
C	4
D	4
E	2
F	6
G	2
H	4

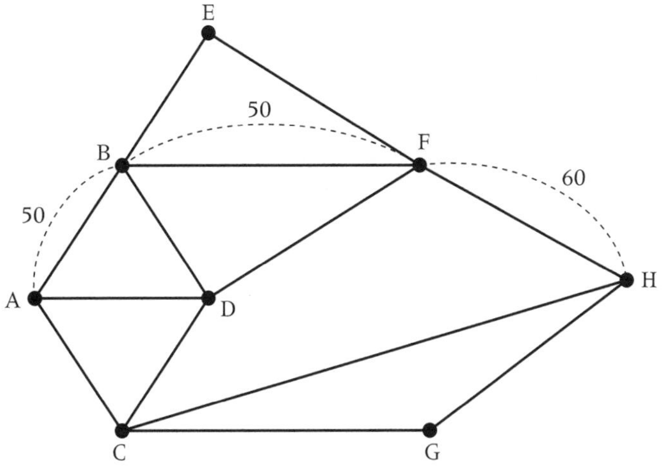

This indicates that the number of times each vertex will appear in the Chinese postman route is:

A $\frac{4}{2} = 2 + 1 = 3$
B $\frac{6}{2} = 3$
C $\frac{4}{2} = 2$
D $\frac{4}{2} = 2$
E $\frac{2}{2} = 1$
F $\frac{6}{2} = 3$
G $\frac{2}{2} = 1$
H $\frac{4}{2} = 2$

The number of vertices in the optimal Chinese postman route is 17. They may be in a different order than in the example above but they must have the number of vertices as indicated in the table.

EXERCISE 3C

Find a route corresponding to an optimal Chinese postman route for the questions in Exercise 3B.

3.6 Variations of the Chinese postman problem

Occasionally problems may be set where the start vertex and the finish vertex do not have to be the same. Any graph with two odd vertices is semi-Eulerian.
For this type of graph the length of the Chinese postman route is the sum of all the edges of a network.
In a network with four vertices, the graph is semi-Eulerian plus two odd edges. In addition to the start and finish vertices there are two other odd vertices.
The shortest Chinese postman route is the sum of all the edges plus the shortest distance connecting the two remaining odd vertices.

Worked example 3.3

A county council is responsible for maintaining the following network of roads. The number on each edge is the length of the road in miles.

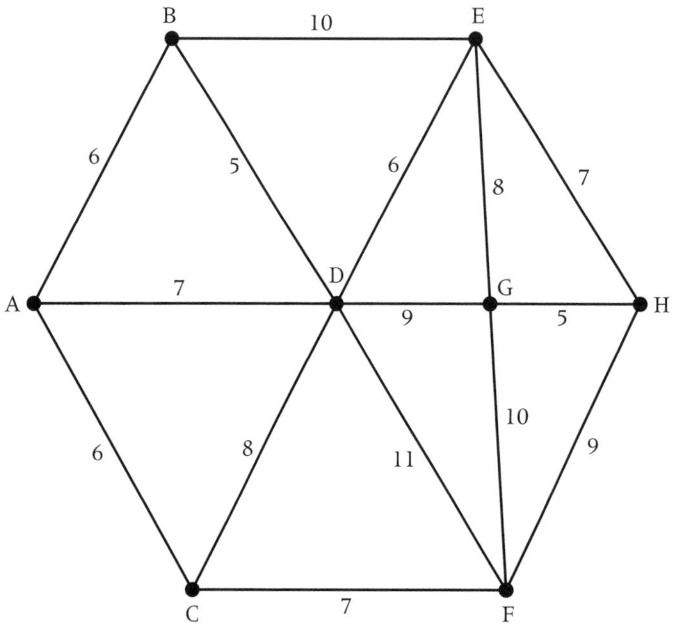

The council office is based at A.

(a) A council worker has to inspect all the roads, starting and finishing at A. Find the length of an optimal Chinese postman route.

(b) A supervisor, based at A, also wishes to inspect all the roads. However, the supervisor lives at H and wishes to start his route at A and finish at H. Find the length of an optimal Chinese postman route for the supervisor.

Solution

(a) There are four odd vertices: A, B, C and H.
There are three ways of pairing these odd vertices and the minimum length of each pairing is:

$$AB + CH = 6 + 16 = 22$$
$$AC + BH = 6 + 17 = 23$$
$$AH + BC = 20 + 12 = 32$$

Draw the edges AB and CH onto the network.
The length of all the roads in the network is 116.
The length of an optimal Chinese postman route for the worker is 114 + 22 = 136 miles.

(b) Starting at A and finishing at H leaves two odd vertices B and C.
The minimum distance from B to C is 12.
The length of an optimal Chinese postman route for the supervisor is 114 + 12 = 126 miles.

EXERCISE 3D

For each of the networks below find the length of an optimal Chinese postman route starting at A and finishing at H.

1

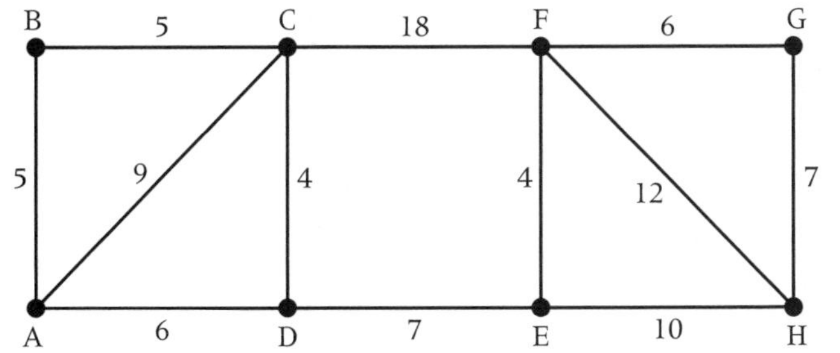

2

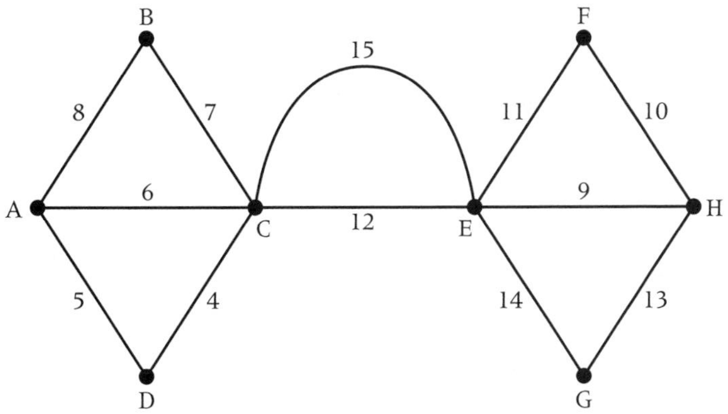

3

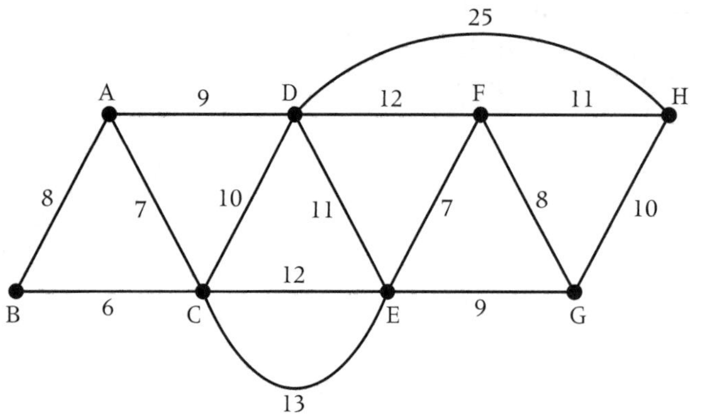

MIXED EXERCISE

1 A local council is responsible for gritting roads.

(a) The following diagram shows the lengths of roads, in miles, that have to be gritted.

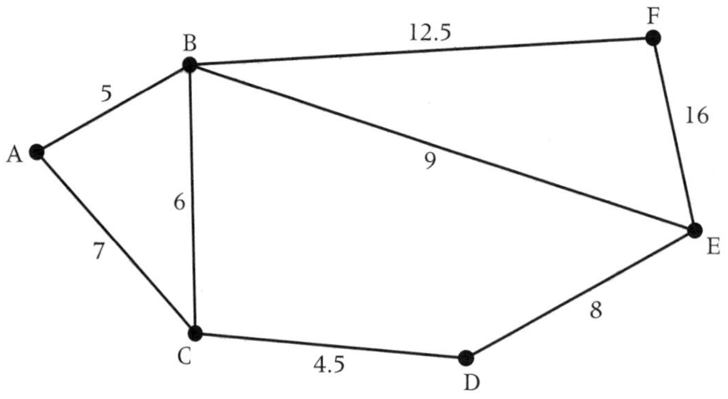

The gritter is based at A and must travel along all the roads, at least once, before returning to A.

(i) Explain why it is **not** possible to start from A and, by travelling along each road only once, return to A.

(ii) Find an optimal Chinese postman route around the network, starting and finishing at A. State the length of your route.

(b) **(i)** The connected graph of the roads in the area run by another council has six odd vertices. Find the number of ways of pairing these odd vertices.

(ii) For a connected graph with n odd vertices, find an expression for the number of ways of pairing these vertices. [A]

2 A road-gritting service is based at a point A. It is responsible for gritting the network of roads shown in the diagram, where the distances shown are in miles.

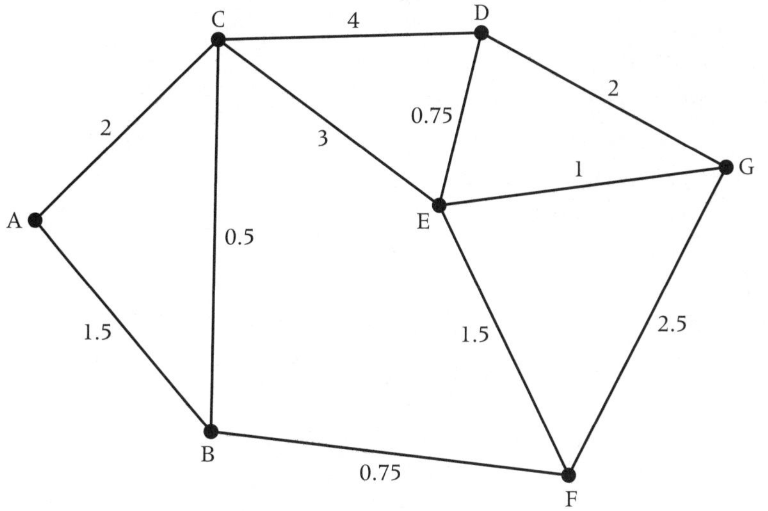

(a) Explain why it is **not** possible to start from A and, by travelling along each road only once, return to A.

(b) In the network there are four odd vertices, B, D, F and G. List the different ways in which these odd vertices can be arranged as two pairs.

(c) For **each** pairing you have listed in **(b)**, write down the sum of the shortest distance between the first pair and the shortest distance between the second pair.

(d) Hence find an optimal Chinese postman route around the network, starting and finishing at A. State the length of your route. [A]

3 A highways department has to inspect its roads for fallen trees.

(a) The following diagram shows the lengths of the roads, in miles, that have to be inspected in one district.

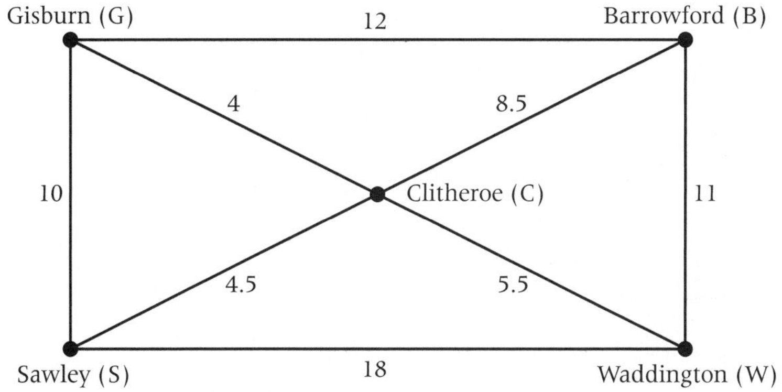

(i) List the three different ways in which the four odd vertices in the diagram can be paired.

(ii) Find the shortest distance that has to be travelled in inspecting all the roads in the district, starting and finishing at the same point.

(b) The connected graph of the roads in another district has six odd vertices. Find the number of ways of pairing these odd vertices.

(c) For a connected graph with n odd vertices, find an expression for the number of ways of pairing these odd vertices. [A]

4 A theme park employs a student to patrol the paths and collect litter. The paths that she has to patrol are shown in the following diagram, where all distances are in metres. The path connecting I and W passes under the bridge which carries the path connecting C and R.

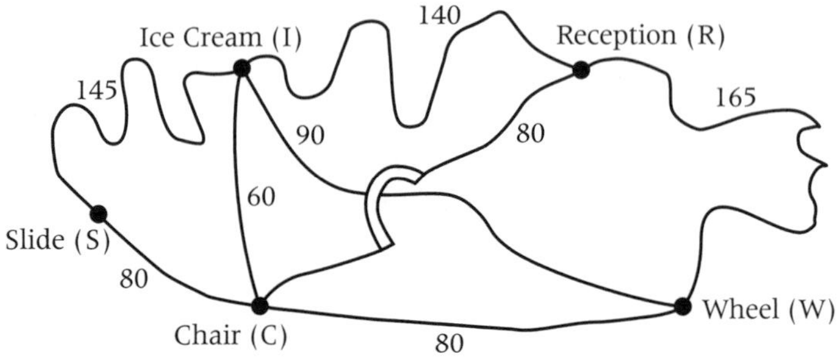

(a) **(i)** Find an optimal Chinese postman route that the student should take if she is to start and finish at Reception (R).

(ii) State the length of your route.

(b) **(i)** A service path is to be constructed. Write down the two places that this path should connect, if the student is to be able to walk along every path without having to walk along any path more than once.

(ii) The distance walked by the student in part **(b)(i)** is shorter than that found in part **(a)(ii)**. Given that the length of the service path is l metres, where l is an integer, find the greatest possible value of l. [A]

5 In the following network the four vertices are odd.

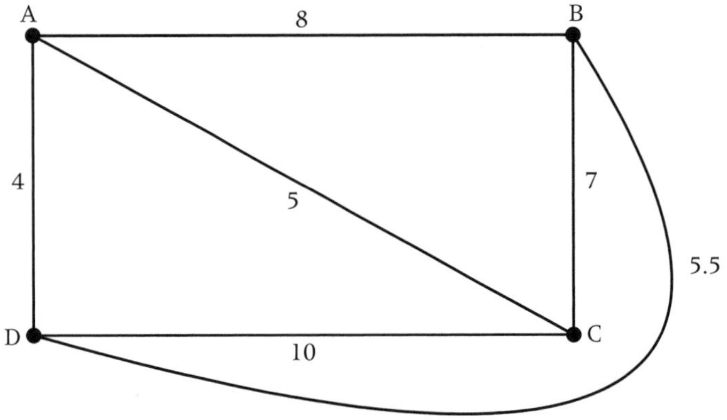

(a) List the different ways in which the vertices can be arranged as two pairs.

(b) For **each** pairing you have listed in **(a)**, write down the sum of the shortest distance between the first pair and the shortest distance between the second pair. Hence find the length of an optimal Chinese postman route around the network.

(c) State the minimum number of extra edges that would need to be added to the network to make the network Eulerian. [A]

3

6 A groundsman at a local sports centre has to mark out the lines of several five-a-side pitches using white paint. He is unsure as to the size of the goal area and he decides to paint the outline as given below, where all the distances are in metres.

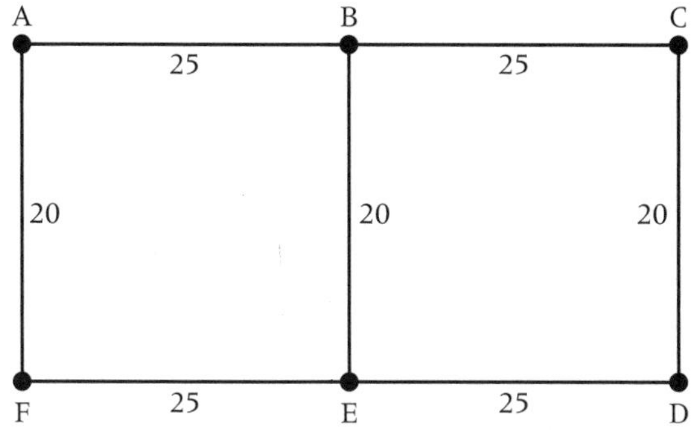

(a) He starts and finishes at the point A. Find the minimum total distance that he must walk and give one of the corresponding possible routes.

(b) Before he starts to paint the second pitch he is told that each goal area is a semi-circle of radius 5 m, as shown in the diagram below.

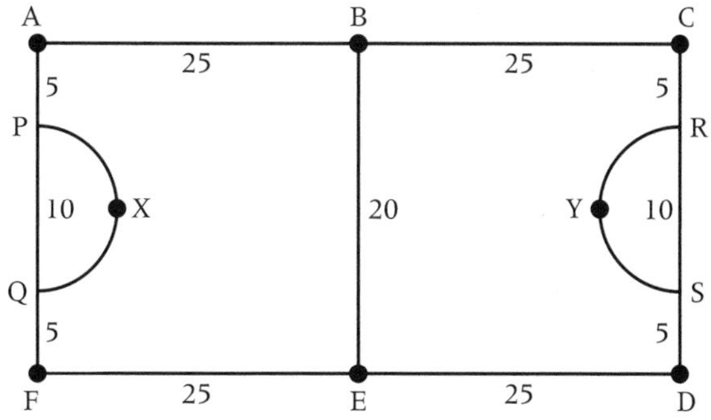

(i) He can start at any point but must return to his starting point. State which vertices would be suitable starting points to keep the total distance walked from when he starts to paint the lines until he completes this task to a minimum.

(ii) Find an optimal Chinese postman route around the lines. Calculate the length of your route. [A]

7 The diagram shows a network of roads connecting five villages. The numbers on the roads are the times, in minutes, taken to travel along each road, where $x > 0.5$.

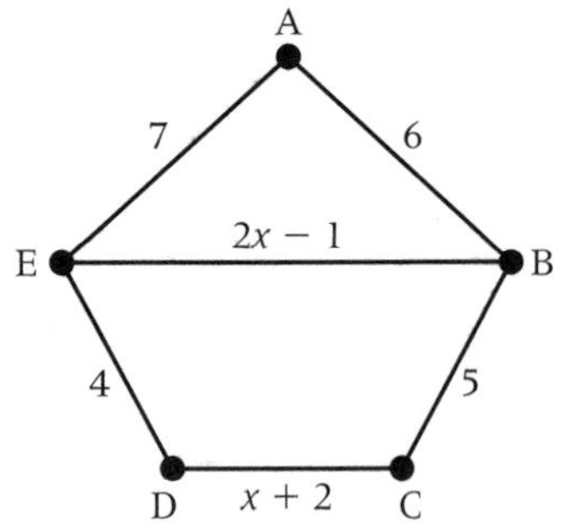

A police patrol car has to travel from its base at B along each road at least once and return to base.

(a) Explain why a route from B to E must be repeated.

(b) List the routes, and their lengths, from B to E, in terms of x where appropriate.

(c) On a particular day, it is known that $x = 10$.
Find the length of an optimal Chinese postman route on this day. State a possible route corresponding to this minimum length.

(d) Find, no matter what the value of x, which of the three routes should **not** be used if the total length of a Chinese postman route is to be optimal. [A]

8 The following question refers to the three graphs: Graph 1, Graph 2 and Graph 3.

Graph 1

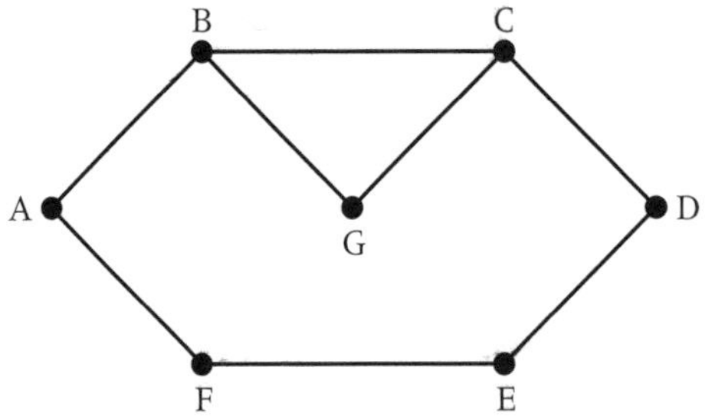

Graph 2

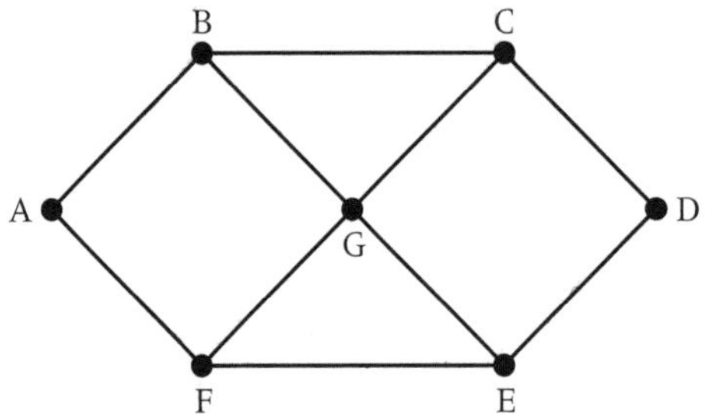

Graph 3

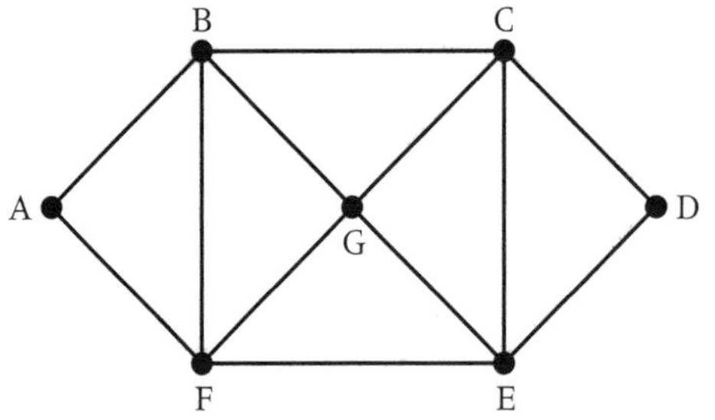

(a) For **each** of the graphs explain whether or not the graph is Eulerian.

(b) The length of each edge connecting two vertices is 1 unit. Find, for **each** of the graphs, the length of an optimal Chinese postman route, starting and finishing at A. [A]

9 The diagram shows the time, in minutes, for a traffic warden to walk along a network of roads, where $x > 0$.

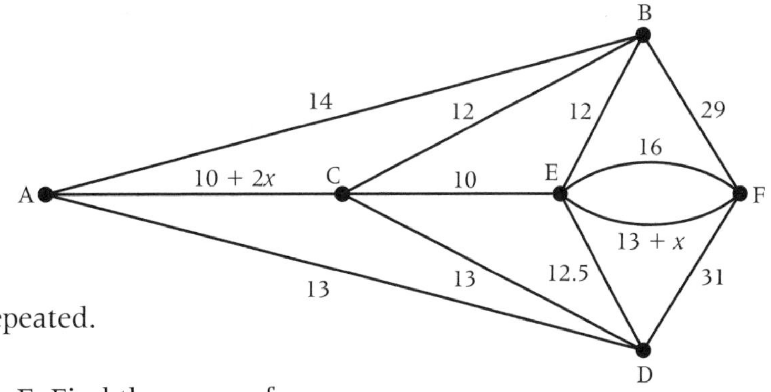

The traffic warden is to start at A and walk along each road at least once before returning to A.

(a) Explain why a section of roads from A to E has to be repeated.

(b) The route ACE is the second shortest route connecting A to E. Find the range of possible values of x.

(c) Find, in terms of x, an expression for the minimum distance that the traffic warden must walk and write down a possible route that he could take.

(d) Starting at A, the traffic warden wants to get to F as quickly as possible. Use Dijkstra's algorithm to find, in terms of x, the minimum time for this journey, stating the route that he should take. [A]

10 The following diagram shows a network of roads connecting six towns. The number on each arc represents the distance, in miles, between towns. The road connecting towns D and F has length x miles, where $x < 13$.

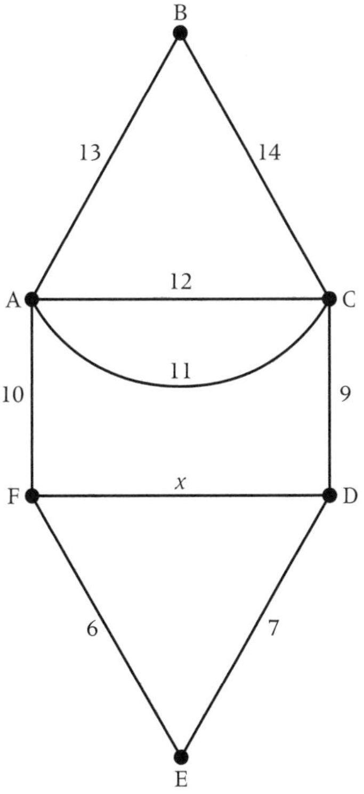

An optimal Chinese postman route, starting and finishing at A, has length 100 miles. Find the value of x. [A]

3

11 The following network shows the distances, in kilometres, of roads connecting ten towns.

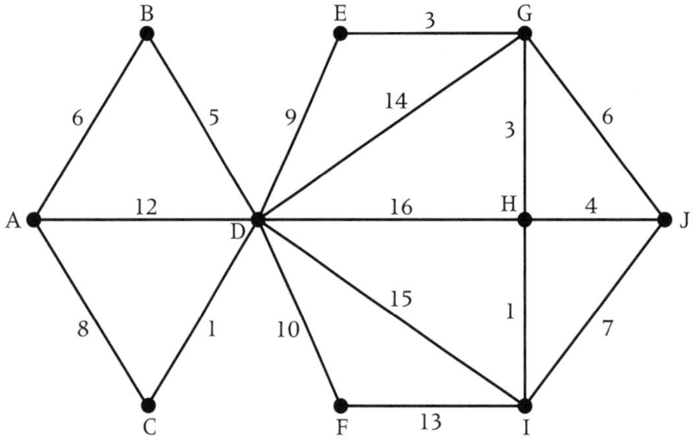

(a) An ambulance is based at A and has to respond to an emergency at J. Use Dijkstra's algorithm to find the minimum distance required to travel from A to J, and state the route.

(b) A police motorcyclist, based at town A, has to travel along each of the roads at least once before returning to base at A. Find the minimum total distance the motorcyclist must travel. [A]

12 The following diagram shows the lengths of roads, in miles, connecting nine towns.

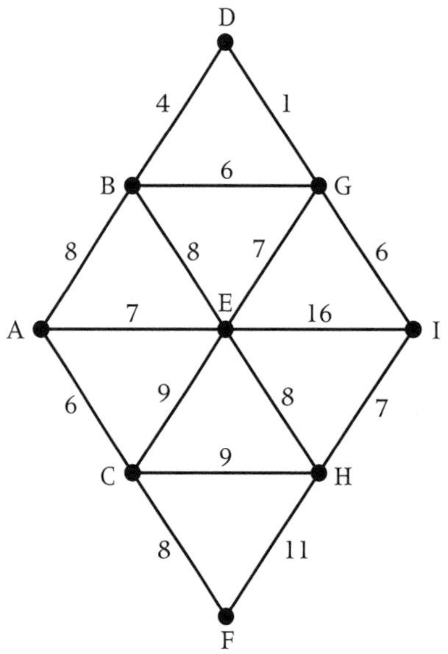

(a) Use Prim's algorithm, starting from A, showing your working at each stage, to find the minimum spanning tree for the network. State its length.

(b) (i) Find an optimal Chinese postman route around the network, starting and finishing at A. You may find the shortest distance between any two towns by inspection.

(ii) State the length of your route. [A]

13 The network on the right has 16 vertices.

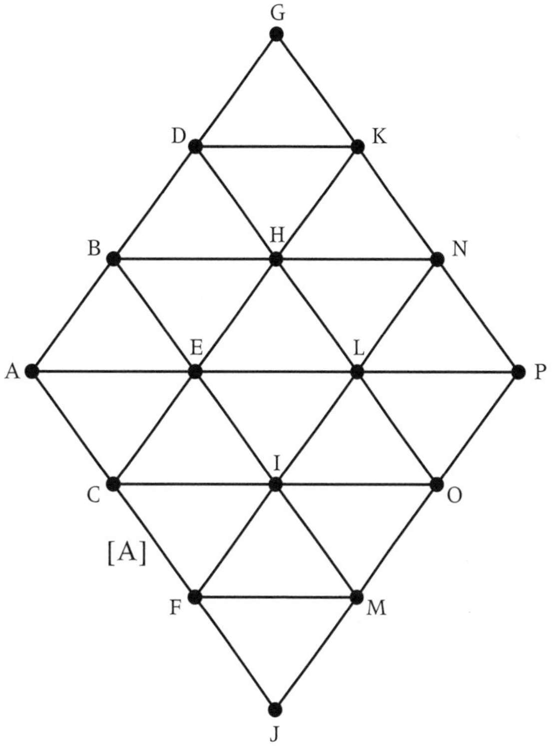

(a) Given that the length of each edge is 1 unit, find:

(i) the shortest distance from A to K

(ii) the length of a minimum spanning tree.

(b) (i) Find the length of an optimal Chinese postman route, starting and finishing at A.

(ii) For such a route, state the edges that would have to be used twice.

(iii) Given that the edges AE and LP are now removed, find the new length of an optimal Chinese postman route, starting and finishing at A. [A]

14 The numbers of parking meters on the roads in a town centre are shown in the network on the right.

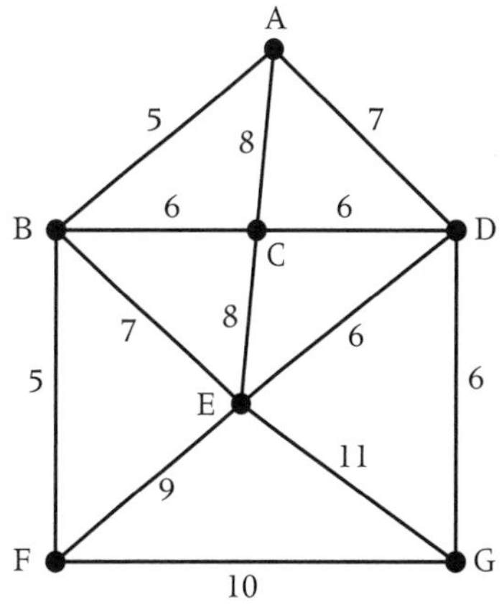

(a) A traffic warden wants to start at A, walk along the roads passing each meter at least once and finish back at A. She wishes to choose her route in order to minimise the number of meters that she passes more than once.

(i) Explain how you know that it will be necessary for her to pass some meters more than once.

(ii) Apply the Chinese postman algorithm to find the minimum number of meters which she will have to pass more than once, and give an example of a suitable route.

(b) At each of the junctions A, B, C, D, E, F and G there is a set of traffic lights. The traffic warden is asked to make a journey, starting and finishing at A, to check that each set of traffic lights is working correctly. Find a suitable route for her which passes 50 or fewer meters. [A]

15 The vertices of the following network represent the chalets in a small holiday park and the arcs represent the paths between them, with the lengths of the paths given in metres.

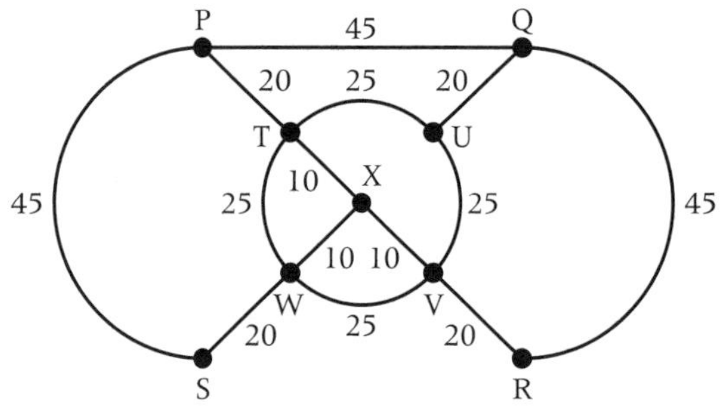

A gardener wishes to sweep all the paths, starting and finishing at P, and to do so by walking (always on the paths) as short a distance as possible. Apply the Chinese postman algorithm to find the shortest distance the gardener must walk, and give one possible shortest route. [A]

Key point summary

1 A **traversable** graph is one that can be drawn without taking a pen from the paper and without retracing the same edge. In such a case the graph is said to have an Eulerian trail. *p45*

2 An **Eulerian trail** uses all the edges of a graph. For a graph to be Eulerian all the vertices must be of even order. *p46*

3 If a graph has two odd vertices then the graph is said to be **semi-Eulerian**. A trail can be drawn starting at one of the odd vertices and finishing at the other odd vertex. *p46*

4 A minimum Chinese postman route requires each edge to be walked along at least once and in addition the least pairings of odd vertices must be walked along on one extra occasion. *p48*

Test yourself | What to review

1 Which of the following networks is traversable? — *Section 3.2*

(a)

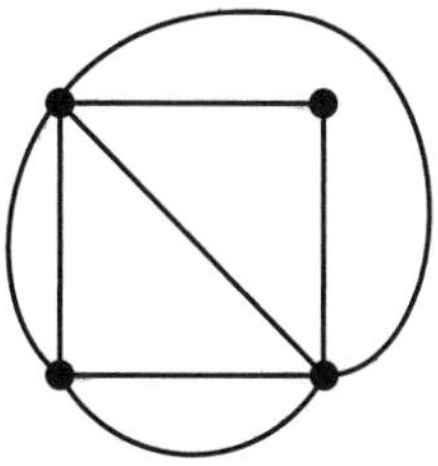

(b)

(c)

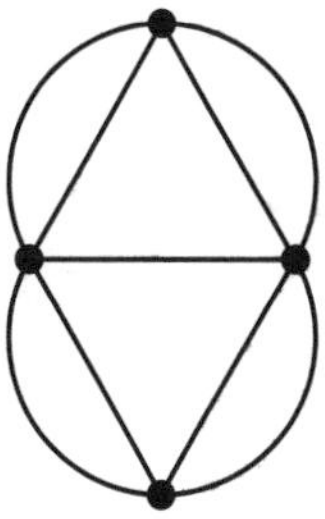

---

2 Find the number of ways of pairing: — *Section 3.3*

(a) 8 odd vertices,

(b) 12 odd vertices,

(c) 20 odd vertices.

---

3 List the ways of pairing the odd vertices in the following networks. For each pairing find the minimum connector. Find the length of an optimal Chinese postman route. Write down one possible route. — *Sections 3.3, 3.4, 3.5*

(a)

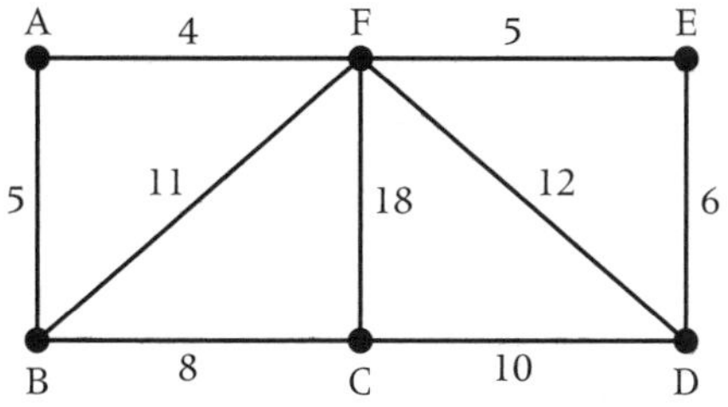

(b)

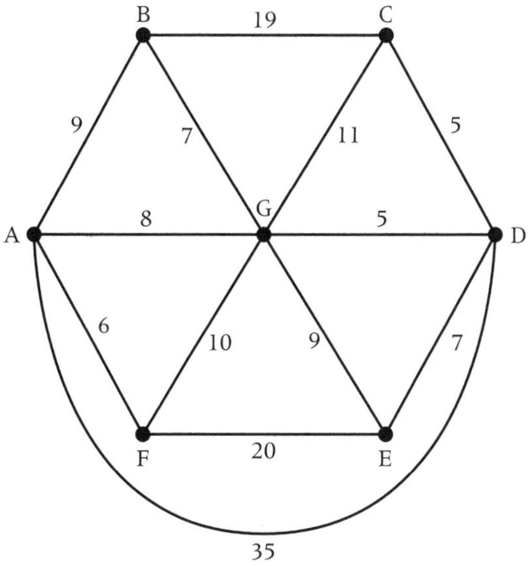

Test yourself ANSWERS

1 **(a)** Yes **(b)** Yes **(c)** No

2 **(a)** 105 **(b)** 10 395 **(c)** 654 729 075

3 **(a)** BC + FD = 19
BD + FC = 35
BF + DC = 19
Total 79 + 19 = 98
AFEDFEDCFBCBA

(b) BC + EF = 36
BE + CF = 36
BF + CE = 27
Total 151 + 27 = 178
ABCDGCDEGBAGFADEFA

CHAPTER 4

Travelling salesman problem

4

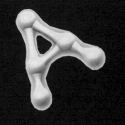

Learning objectives

After studying this chapter, you should be able to:

- understand the travelling salesman problem
- solve it for simple networks
- find upper and lower bounds for the problem
- convert a practical problem into a classical problem, solve and interpret the answer.

4.1 Introduction

Chapter 3 was based on finding routes around a network where every edge was used at least once. This chapter investigates tours around a network in which every vertex is visited at least once. This is called the travelling salesman problem.

A tour is dealt with fully in Chapter 5.

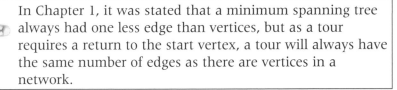

In Chapter 1, it was stated that a minimum spanning tree always had one less edge than vertices, but as a tour requires a return to the start vertex, a tour will always have the same number of edges as there are vertices in a network.

This is the type of situation in which a company representative has to visit a number of stores before returning to base.

There are many similar practical situations where this type of problem occurs: a politician touring a number of towns on election day, a delivery van making deliveries to a number of shops, a shopper comparing prices at a number of stores before making a purchase.

In each situation it is important to minimise the length of the tour.

(The word 'length' may be used to describe distance, time or money.)

Currently there is no algorithm for finding a tour of minimum length, apart from a complete enumeration, but, conversely, it has not been proven that such an algorithm does not exist. A complete enumeration is impractical in all but the simplest of networks.

Worked example 4.1

Find the number of different tours for the complete network of four vertices shown below.

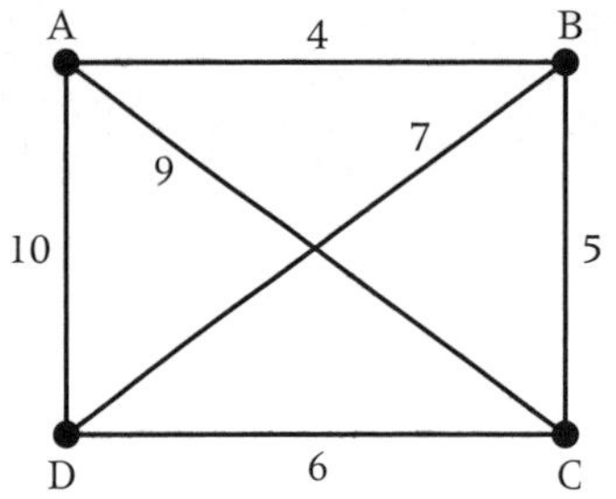

Solution

There are three distinct tours:

ABCDA of length 25
ABDCA of length 26
ACBDA of length 31

There are of course three other tours, ADCBA, ACDBA and ADBCA, but these tours are the same as the first three in reverse order.

Although there are three distinct tours in this case, we can't assume that tours in reverse order are of the same length.

Worked example 4.2

The diagram below shows a one-way road system connecting four points. Find the number of distinct tours.

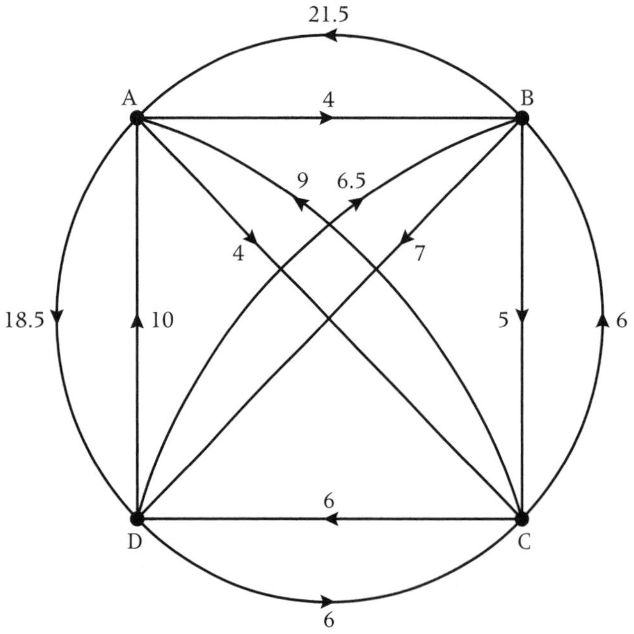

Solution

There are six distinct tours:

ABCDA of length 25
ABDCA of length 26
ACBDA of length 27
ACDBA of length 38
ADBCA of length 39
ADCBA of length 52

In each tour there are four numbers to add; hence the total number of calculations is $4 \times 3 \times 2 \times 1 = 24$, which is 4!.
In general, for an undirected network consisting of n vertices, there are $(n - 1)!/2$ tours, and $n!/2$ calculations.
For a directed network there are $(n - 1)!$ tours and $n!$ calculations.
The following table shows the number of possible tours and number of calculations for an undirected network consisting of n vertices.

Number of vertices	Number of tours	Number of calculations
3	1	3
4	3	12
5	12	60
10	181 440	1 814 400
20	6.08×10^{16}	1.22×10^{18}

If a computer with a 2400 MHz processor were to check all possible tours for a network with only 10 vertices, it would take 12.6 minutes!
As a complete enumeration is impractical and there is no algorithm for finding an optimum solution, how do we proceed?
There are algorithms for finding ranges of possible solutions and we will investigate their applications and implications.

4.2 Upper bound

An upper bound is defined as being a tour that may be improved upon. This is very significant.

Any tour that exists is an **upper bound**.

For a complete graph, any random order of the vertices will produce a tour and hence an initial upper bound, but is there a systematic approach for finding an upper bound?
The problem we are posed with is: how do we find a best upper bound?

Worked example 4.3

Students try different orders of vertices to try to find a tour of minimum length. Their results are 46, 42, 45, 50, 43, 45, 41 and 48. Write down the best upper bound and explain why your answer is the best upper bound.

Solution

The best upper bound is 41.
Although all other values give tours, they all can be improved on. From the figures given we know that 41 exists, but we haven't an improved result.

> The best upper bound is the **lowest upper bound**.

4.3 Nearest-neighbour algorithm

A logical method for finding an upper bound of a complete network is to always choose to go to the nearest unvisited vertex at each stage. The following is a simple set of instructions for obtaining an upper bound.

> Nearest-neighbour algorithm
>
> **Step 1** Choose a start vertex.
>
> **Step 2** From your current vertex go to the nearest unvisited vertex.
>
> **Step 3** Repeat step 2 until all the vertices have been visited.
>
> **Step 4** Return to the start vertex.

The algorithm can be applied to either a network diagram or directly to a table.

Worked example 4.4

Use the nearest-neighbour algorithm, from each vertex, to find three upper bounds for the network shown.
State the best upper bound.

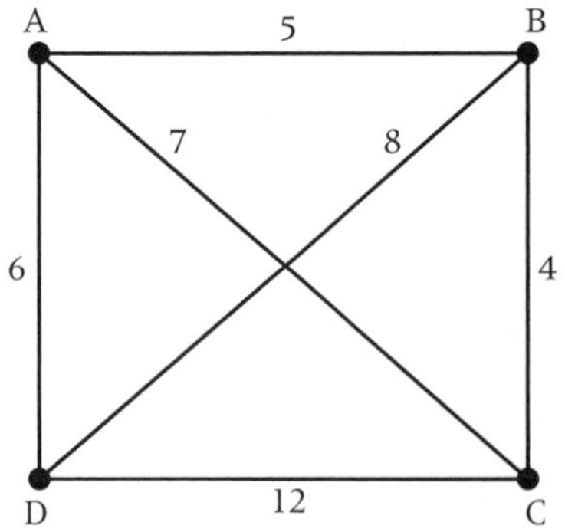

Solution

Step 1 Start from A.

Step 2 From A, the nearest vertex is B, with length 5.

Step 2 From B, the nearest vertex is C, with length 4.

Step 2 From C, the nearest unvisited vertex is D, with length 12.

Step 3 All vertices have been visited.

Step 4 Return to A, length 6.

Tour is ABCDA, with length 27.

Step 1 Start from B.

Step 2 From B, the nearest vertex is C, with length 4.

Step 2 From C, the nearest unvisited vertex is A, with length 7.

Step 2 From A, the nearest unvisited vertex is D, with length 6.

Step 3 All vertices have been visited.

Step 4 Return to B, length 8.

Tour is BCADB, with length 25.

Step 1 Start from C.

Step 2 From C, the nearest vertex is B, with length 4.

Step 2 From B, the nearest unvisited vertex is A, with length 5.

Step 2 From A, the nearest unvisited vertex is D, with length 6.

Step 3 All vertices have been visited.

Step 4 Return to C, length 12.

Tour is CBADC, with length 27.

Step 1 Start from D.

Step 2 From D, the nearest vertex is A, with length 6.

Step 2 From A, the nearest unvisited vertex is B, with length 5.

Step 2 From B, the nearest unvisited vertex is C, with length 4.

Step 3 All vertices have been visited.

Step 4 Return to D, length 12.

Tour is DABCD, with length 27.
The four upper bounds have lengths 27, 25, 27 and 27.
The best upper bound is 25.
Why?
A tour of length 25 is possible and although it may be improved on, it is lower than the other upper bounds.

EXERCISE 4A

1 Use the nearest-neighbour algorithm starting from each vertex in turn to find upper bounds for the following networks. In each case state the best upper bound and the corresponding route.

(a)

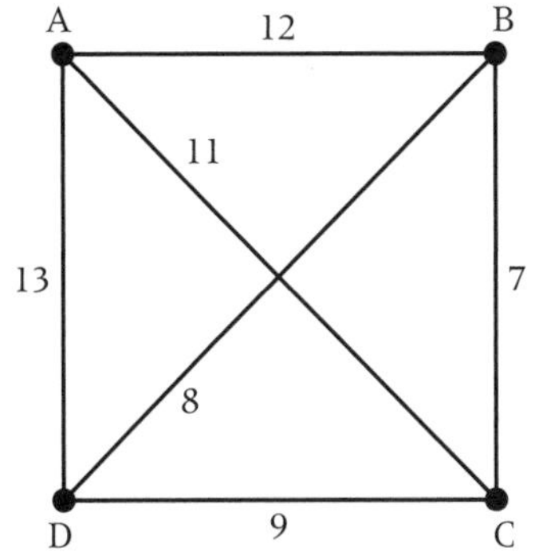

(b)

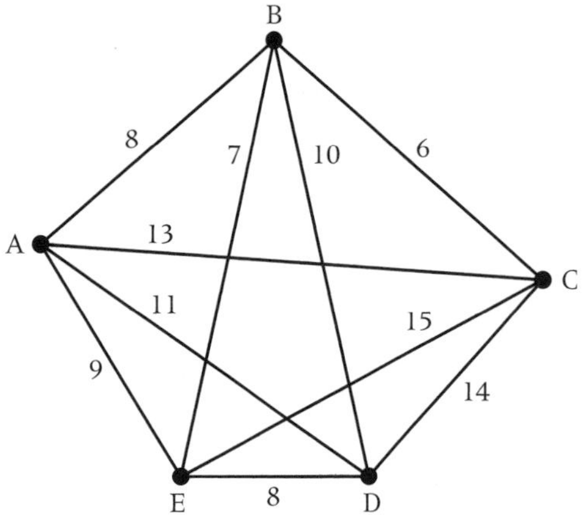

(c)

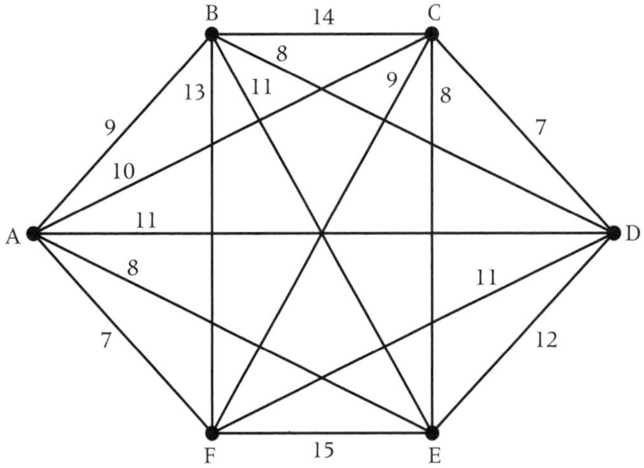

2 Use the nearest-neighbour algorithm starting from each vertex in turn to find upper bounds for the following networks. In each case state the best upper bound and the route corresponding to this upper bound.

(a)

	A	**B**	**C**
A	–	12	9
B	12	–	13
C	9	13	–

(b)

	A	**B**	**C**	**D**	**E**
A	–	8	5	7	3
B	8	–	7	4	6
C	5	7	–	10	8
D	7	4	10	–	11
E	3	6	8	11	–

(c)

To \ From	A	B	C	D	E	F
A	–	12	8	13	9	11
B	11	–	9	12	14	10
C	12	13	–	8	7	9
D	10	9	11	–	11	13
E	8	8	13	9	–	14
F	14	15	12	8	11	–

4.4 Limitations of the nearest-neighbour algorithm

The algorithm always selects the nearest unvisited vertex, it does not consider the implications of the selection. It may be obvious by inspection which edge of a network you would not want to include in a tour, but the edge may be included by following the algorithm.

Worked example 4.5

Use the nearest-neighbour algorithm for the following network, starting from A, to find an upper bound for the length of a minimum tour.

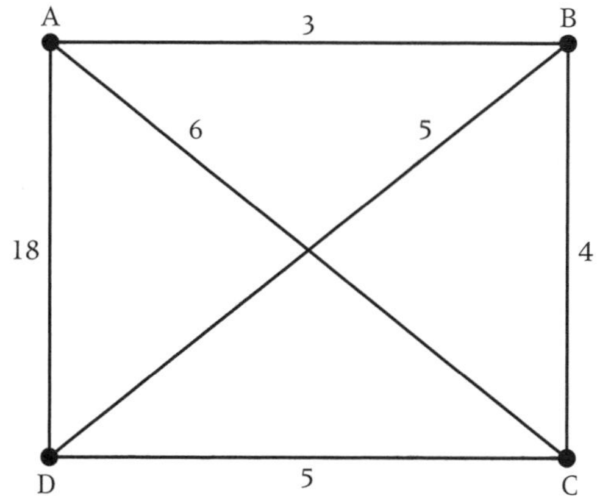

Solution

Starting from A, the algorithm gives the tour of ABCDA, which has a length of 30.
However, it is obvious from the diagram that the edge of AD should not be included in a minimum tour. From A, the algorithm selects the vertices B then C, leaving D as the last vertex selected. From D, the edge DA has to be included to return to the start vertex.

In this case, the minimum length of a tour is 19 and the corresponding route is ABDCA.

Worked example 4.6

The following network shows roads connecting six villages. Three villages are on one side of a river and are connected by a bridge at A to the other three villages.
Use the nearest-neighbour algorithm starting from A to find an upper bound for the length of a minimum tour.

The algorithm is only valid for complete networks.

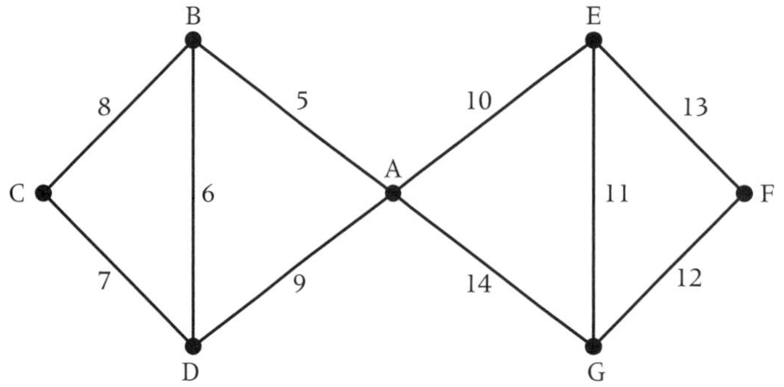

Solution

From A the nearest vertex is B.
From B the nearest vertex is D.
From D the nearest vertex is C.
There is now a problem because, from C there are still three vertices to visit, but C is only connected to B and D.
In this situation vertices need to be revisited and as such the algorithm fails!
This problem occurred because the algorithm is only valid for complete networks (see Chapter 5, page 93).

4.5 Lower bound

A lower bound is defined as being the lowest possible value for a tour, but the tour may not exist.
This is a difficult concept.
A value can be calculated that we know the length of a tour cannot be less than; however, although we have calculated a value this doesn't guarantee that such a tour is possible.
When you studied upper bounds earlier in this chapter, you learnt that the best upper bound is the lowest upper bound; the reverse is true for lower bounds.

> The best lower bound is the greatest lower bound.

In section 4.1 you learnt that any tour would have the same number of edges as vertices. If you choose minimum edges then you can be certain that a tour cannot be lower than this value.
How do we choose the minimum edges?

In chapter 1 you used algorithms to find a minimum spanning tree for a network, so this must be our starting point.

Lower bound algorithm

Step 1 Delete a vertex and all edges connected to the vertex.

Step 2 Find a minimum spanning tree for the remaining network.

Step 3 Add the two shortest edges from the deleted vertex.

Use either Prim's or Kruskal's algorithm to find the minimum spanning tree.

4

If the algorithm is applied to each vertex in turn then the best lower bound is the largest of the answers.

Worked example 4.7

By deleting each vertex in turn, find four lower bounds for the length of a minimum tour, and hence state the best lower bound.

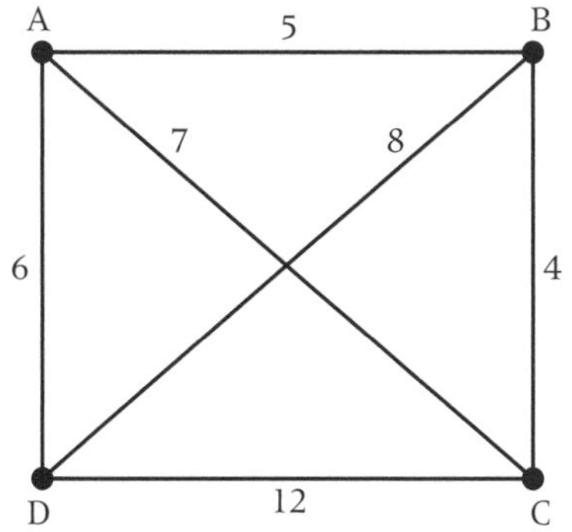

Solution

Delete A and find the minimum spanning tree for the other vertices. The minimum spanning tree is:

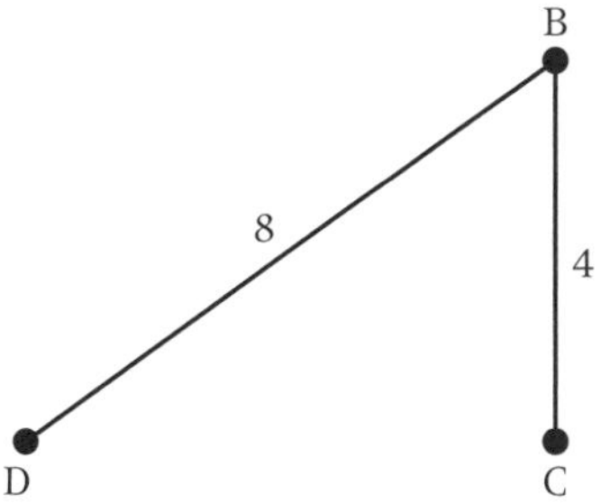

The length of the minimum spanning tree is 12.
The two shortest edges from A are 5 and 6.
Hence a lower bound is $12 + 5 + 6 = 23$.

Delete B and the minimum spanning tree for the other vertices is:

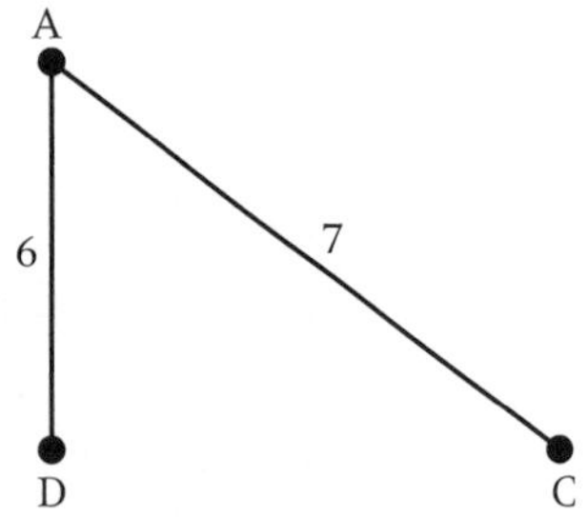

The length of the minimum spanning tree is 13.
The two shortest edges from B are 4 and 5.
Hence a lower bound is $13 + 4 + 5 = 22$.

Delete C and the minimum spanning tree for the other vertices is:

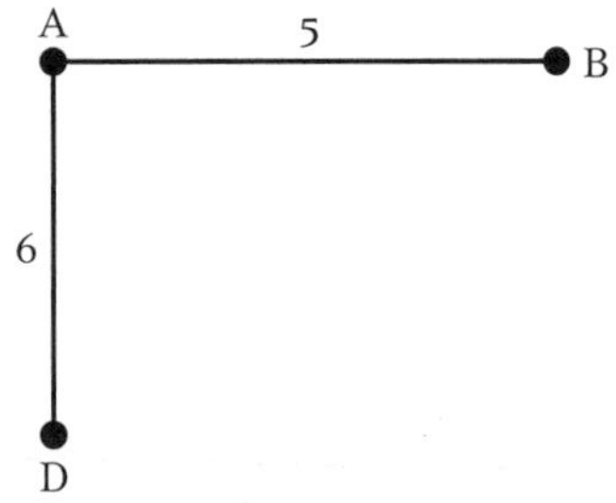

The length of the minimum spanning tree is 11.
The two shortest edges from C are 4 and 7.
Hence a lower bound is $11 + 4 + 7 = 22$.

Delete D and the minimum spanning tree for the other vertices is:

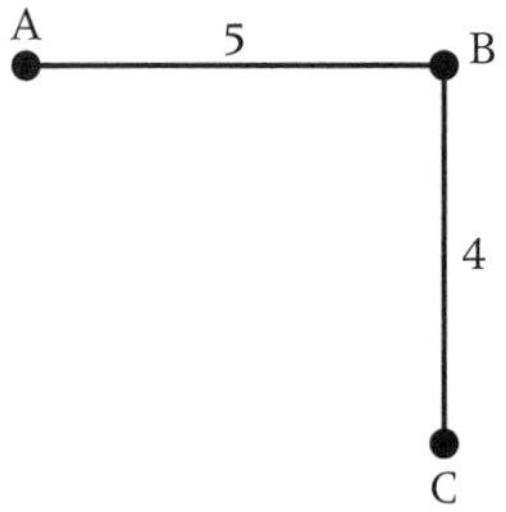

The length of the minimum spanning tree is 9.
The two shortest edges from A are 6 and 8.
Hence a lower bound is $9 + 6 + 8 = 23$.

The four lower bounds are 22, 22, 22 and 23.
The best lower bound is 23. We know that it is impossible to have a tour of less than 23.

When you were working on upper bounds for this problem, you found that the best upper bound was 25.
You can now put the two answers together by writing

$$23 \leqslant \text{minimum tour} \leqslant 25$$

EXERCISE 4B

1 By deleting each vertex in turn, find lower bounds for the following networks. In each case state the best lower bound.

(a)

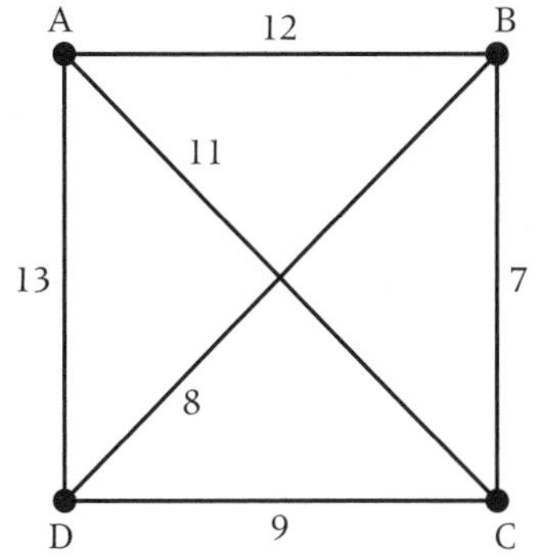

(b)

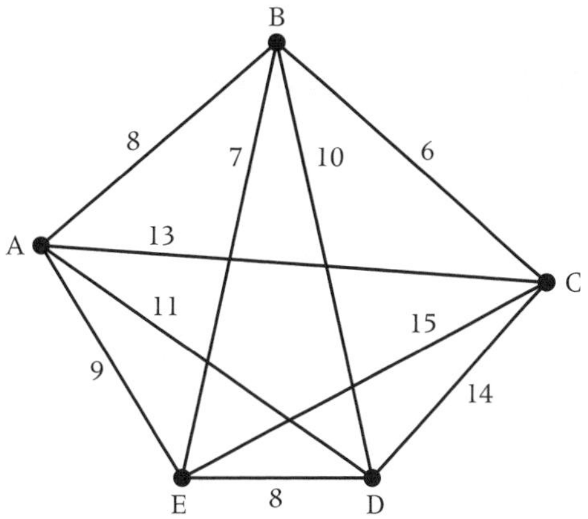

(c)

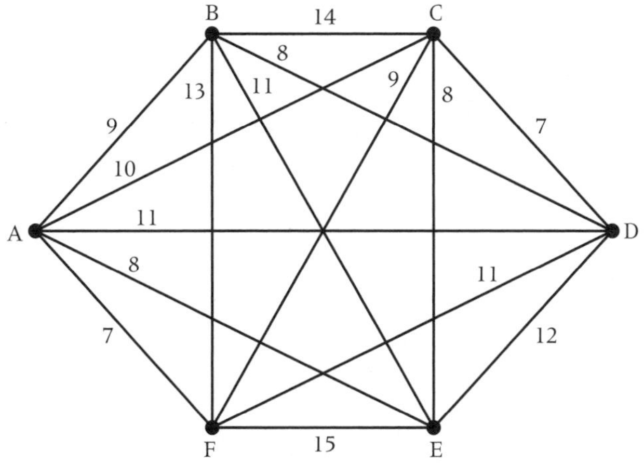

2 By deleting each vertex in turn find lower bounds for the following networks. In each case state the best lower bound.

(a)

	A	B	C
A	–	12	9
B	12	–	13
C	9	13	–

(b)

	A	B	C	D	E
A	–	8	5	7	3
B	8	–	7	4	6
C	5	7	–	10	8
D	7	4	10	–	11
E	3	6	8	11	–

(c)

To \ From	A	B	C	D	E	F
A	–	12	8	13	9	11
B	11	–	9	12	14	10
C	12	13	–	8	7	9
D	10	9	11	–	11	13
E	8	8	13	9	–	14
F	14	15	12	8	11	–

3 By considering your answers to questions 1 and 2 in Exercises 4A and 4B, write down the smallest interval within which the length of the minimum tour must lie.

4.6 Incomplete networks

Worked example 4.8

Find an upper bound for the length of a minimum tour for the following network by visiting the vertices in the following order: ABCDEFCGA.
By deleting C, find a lower bound for the length of a tour.

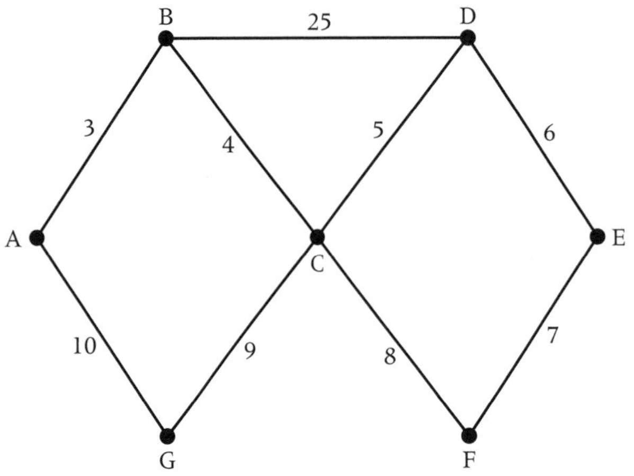

Solution

The tour ABCDEFCGA is of length
$3 + 4 + 5 + 6 + 7 + 8 + 9 + 10 = 52$.
As this tour exists, it must be an upper bound.
To find a lower bound by deleting C, we obtain the following minimum spanning tree.

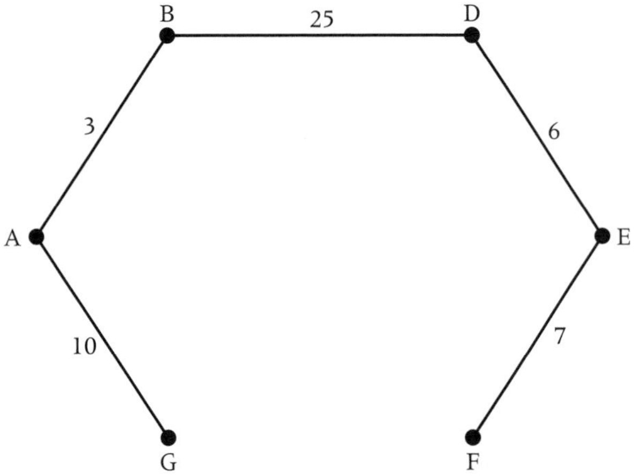

The length of this spanning tree is 51.

Add the two shortest edges from C, BC and CD of lengths 4 and 5.
This gives a lower bound of 60.

Conclusion: the upper bound is 52 and the lower bound is 60!

The implication of this statement is that we can definitely complete a tour in 52, but this figure may be lower, and we definitely cannot complete a tour in 60, but this figure may be higher!

$$60 \leqslant \text{minimum tour} \leqslant 52$$

This ridiculous result has occurred because the original network was incomplete. Also the triangle BCD did not obey the triangle inequality.

> If any network is incomplete then before any upper or lower bounds are obtained, the network must be made complete. Ensure that the distances between all pairs of vertices are represented by a single edge of minimum length.

Worked example 4.9

Let us now revisit the same network.

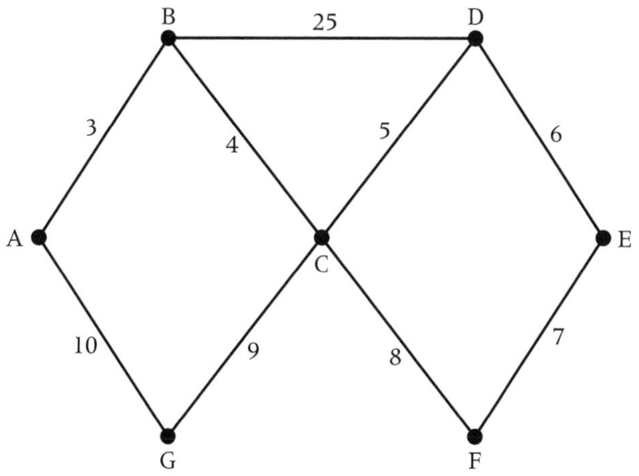

Solution

We find the minimum distance between pairs of vertices to be:

AB = 3
AC = 7 (AB + BC)
AD = 12 (AB + BC + CD)
AE = 18 (AB + BC + CD + DE)
AF = 15 (AB + BC + CF)
AG = 10

Repeat the process from the other vertices.

The complete network can be summarised by the table below.

	A	B	C	D	E	F	G
A	–	3	7	12	18	15	10
B	3	–	4	9	15	12	13
C	7	4	–	5	11	8	9
D	12	9	5	–	6	13	14
E	18	15	11	6	–	7	20
F	15	12	8	13	7	–	17
G	10	13	9	14	20	17	–

An upper bound ABCDEFGA has length
$3 + 4 + 5 + 6 + 7 + 17 + 10 = 52$. This is the same result as earlier.
To find a lower bound by deleting C, we obtain the following minimum spanning tree:

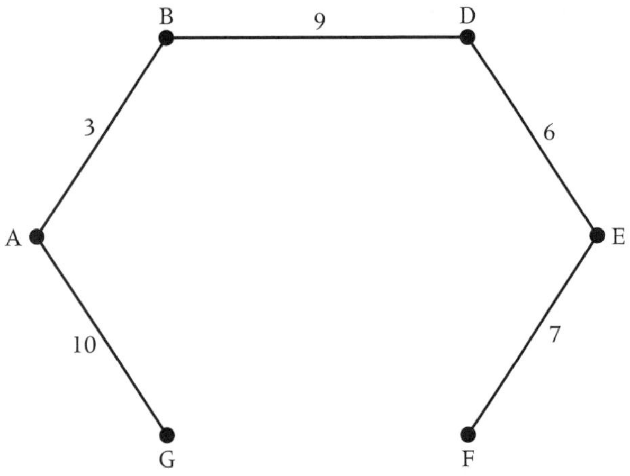

The length of this minimum spanning tree is 35.
The two shortest edges from C are 4 and 5.
The lower bound is $35 + 9 = 47$.
Hence

$$47 \leqslant \text{minimum tour} \leqslant 52$$

The actual answer for a minimum tour is 52.

We have solved the classical problem of finding a tour of minimum length for a complete network. We have to now consider the implications for the original incomplete network.

The classical answer is a tour ABCDEFGA, but the distance of 17 from F to G is found by revisiting C. We must replace FG in the tour by FCG.
The solution to the original problem is ABCDEFCGA with length 52.

EXERCISE 4C

For each of the networks below:

(a) make the network complete,
(b) for the complete network find upper and lower bounds by considering each vertex in turn,
(c) write down the route of the complete network corresponding to the best upper bound,
(d) write down this route in relation to the original network.

1

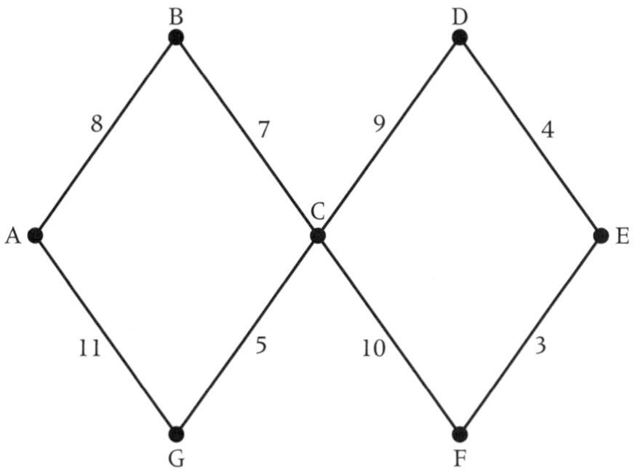

2

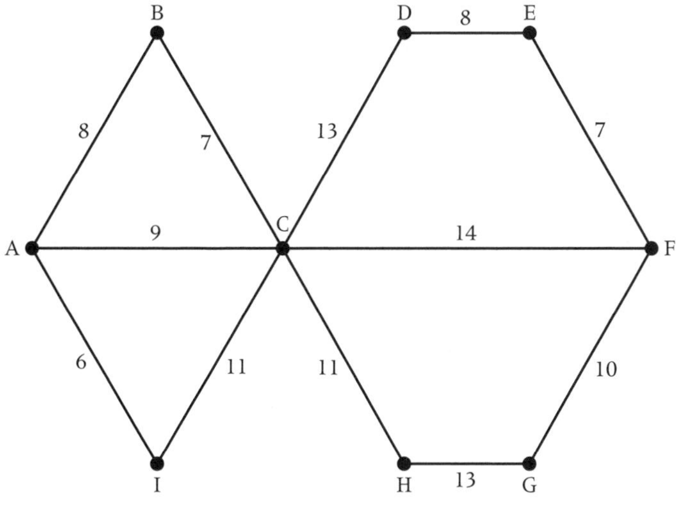

3

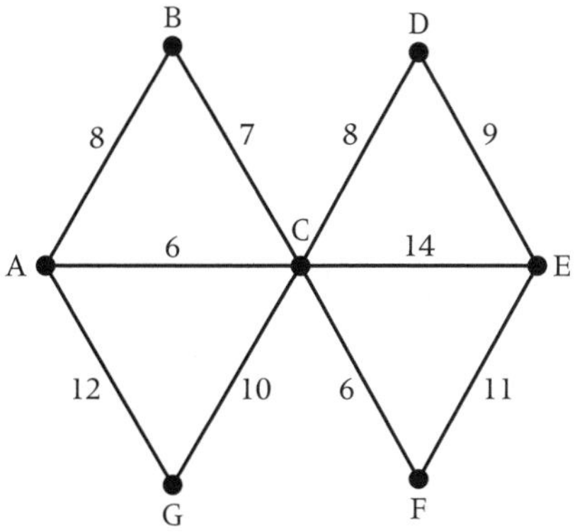

MIXED EXERCISE

1 A company based in Rochdale produces a free newspaper for distribution locally. The following table shows the six surrounding towns that receive the free paper. The figures represent the time, in minutes, to travel between the towns. The company delivery van has to travel from Rochdale to each one of the other towns, before returning to Rochdale.

	Rochdale	Castleton	Middleton	Shaw	Milnrow	Littleborough	Whitworth
Rochdale	–	3	7	8	6	5	4
Castleton	3	–	9	6	8	7.5	6.5
Middleton (Mn)	7	9	–	14	12	11.5	12
Shaw	8	6	14	–	13	12	11
Milnrow (Mi)	6	8	12	13	–	10	9
Littleborough	5	7.5	11.5	12	10	–	8
Whitworth	4	6.5	12	11	9	8	–

(a) Find a minimum connector for the seven towns, stating its length.

(b) Use the nearest-neighbour algorithm, starting from Rochdale, to find an upper bound for a tour of the seven towns.

(c) By deleting Rochdale from the minimum connector found in **(a)**, obtain a lower bound for a tour of the seven towns. [A]

2 Joanne, who lives in Rochdale, decides that she wants to visit all the universities during a weekend. The following table shows the distances, in miles, between Rochdale and the universities, and between the universities themselves. Joanne is going to travel from Rochdale, visiting each of the five universities before returning to Rochdale.

	Rochdale	Bristol	Cambridge	Durham	Leeds	Oxford
Rochdale	–	186	143	125	38	162
Bristol	186	–	128	255	192	71
Cambridge	143	128	–	188	129	68
Durham	125	255	188	–	74	225
Leeds	38	192	129	74	–	148
Oxford	162	71	68	225	148	–

(a) Use the nearest-neighbour algorithm, starting from Rochdale, to find an upper bound for Joanne's tour.

(b) By deleting Rochdale find a lower bound for Joanne's tour. [A]

3 The island of Lanzarote is short of water and the local government is to implement a new system so that water is distributed to all the main towns on the island. The main source of water is at Arrecife (A). The diagram shows the main towns and the distances, in kilometres, between them.

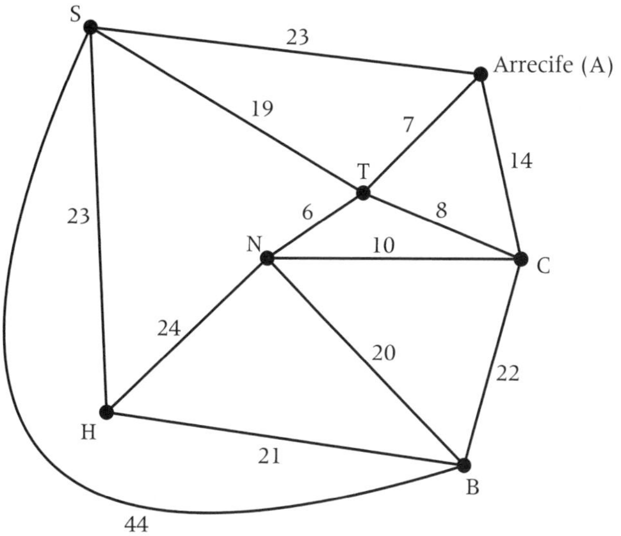

(a) Using Prim's algorithm starting from A, and showing your working at each stage, find the minimum length of piping needed. Show your final network of piping as a minimum spanning tree.

(b) A tourist lands at the airport at Arrecife (A) and is to tour the island, visiting each of the other six towns before returning to A.

(i) By deleting A, find a lower bound for the length of such a tour.

(ii) Explain why a tour of such a length is impossible in this situation. [A]

4 Paul, who lives in Manchester, wants to visit one of the universities in each of Manchester, Liverpool, Sheffield and Birmingham before he fills in his university application form. He decides to spend part of his holiday visiting each university by travelling directly from one to the next. The table below shows the distances, in kilometres, between the universities.

	Manchester	**Liverpool**	**Sheffield**	**Birmingham**
Manchester	–	56	61	129
Liverpool	56	–	116	150
Sheffield	61	116	–	122
Birmingham	129	150	122	–

By deleting each university in turn, find four lower bounds and hence the best lower bound for the distance which Paul must travel between the four universities. [A]

4

5 A message is to be taken by a secretary to each of five classrooms. The secretary will leave the school office and will go to each classroom once before returning to the school office.

The table gives the times, in seconds, taken by the secretary to walk between pairs of rooms.

	Office	**Room 1**	**Room 2**	**Room 3**	**Room 4**	**Room 5**
Office	–	44	52	54	50	48
Room 1	44	–	56	50	54	53
Room 2	52	56	–	66	62	61
Room 3	54	50	66	–	64	62
Room 4	50	54	62	64	–	60
Room 5	48	53	61	62	60	–

(a) **(i)** The secretary leaves the office and visits rooms 1, 2, 3, 4 and 5 in that order before returning to the office. Find the walking time for this tour.

(ii) Explain why this answer may be considered as an upper bound for the minimum total walking time for the secretary's tour.

(b) Use the nearest-neighbour algorithm, starting from the office, to obtain an improved upper bound.

(c) By initially ignoring the office, find a lower bound for the walking time of the secretary's tour. [A]

6 A saleswoman has to visit six towns during the day before returning home. The distances, in miles, between the six towns are given in the following table.

	A	**B**	**C**	**D**	**E**	**F**
A	–	22	26	16	30	26
B	22	–	20	32	10	12
C	26	20	–	34	16	16
D	16	32	34	–	34	32
E	30	10	16	34	–	8
F	26	12	16	32	8	–

(a) Find a minimum spanning tree for the six towns.

(b) To try to find a shorter tour, the saleswoman deletes town E and obtains a lower bound of 84 miles. She obtains the same result by deleting town F.

By deleting the other four towns in turn, find four further lower bounds for the distance she must travel in visiting the six towns. [A]

7 Roger, a football supporter, is to visit each of six football grounds. He decides to travel from one ground to the next until he has visited all of the grounds, starting and finishing at Man. City. The following table shows the distances, in miles, between the grounds.

	Man. City	Burnley	Crewe	Preston	Stockport	Tranmere
Man. City	–	19	26	32	8	31
Burnley	19	–	43	21	22	36
Crewe	26	43	–	42	19	23
Preston	32	21	42	–	36	26
Stockport	8	22	19	36	–	27
Tranmere	31	36	23	26	27	–

(a) Use the nearest-neighbour algorithm, starting and finishing at Man. City, to find an upper bound for the total distance Roger must travel.

(b) By initially ignoring Man. City, find a lower bound for the total distance he must travel in visiting the six grounds.

(c) Using your answers to **(a)** and**(b)**, write down inequalities for D, the total distance, in miles, that Roger has to travel. [A]

8 During a general election campaign a politician, who is based at A, has to visit towns B, C and D on a particular day before returning to A. He is trying to find the tour that will minimise his travelling distance. The following table shows the distances, in kilometres, between the four towns.

From \ To	A	B	C	D
A	–	120	80	60
B	120	–	75	90
C	80	75	–	65
D	60	90	65	–

(a) Use the nearest-neighbour algorithm to find a possible tour and state its length.

(b) Explain why the length of this tour may be considered to be an upper bound. [A]

4

9 A travelling salesman wishes to visit each of eight towns, returning to the town where he started.

The following diagram shows the distances, in miles, between the eight towns.

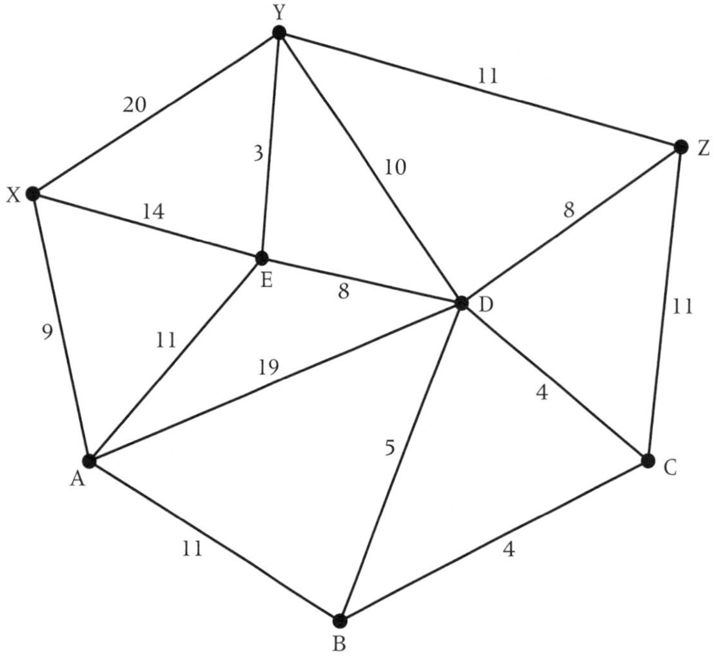

(a) Find a minimum spanning tree for the eight towns, stating its length.

(b) Use your answer to **(a)** to find an upper bound for a tour of the eight towns.

(c) Use the nearest-neighbour algorithm to find an improved upper bound starting:

(i) from town Z,

(ii) from town Y.

(d) By deleting town X obtain a lower bound for a tour.

(e) Hence write down inequalities for L, the minimum length of a tour of the eight towns. [A]

10 A sweet company has a production line making batches of seven different flavours of sweets, A, B, C, D, E, F, G. The changeover times, in minutes, from the production line being set up for one flavour to it being set up for another flavour are given in the following table.

	A	**B**	**C**	**D**	**E**	**F**	**G**
A	–	13	17	18	16	15	14
B	13	–	19	16	18	17.5	16.5
C	17	19	–	24	22	21.5	22
D	18	16	24	–	23	22	21
E	16	18	22	23	–	20	19
F	15	17.5	21.5	22	20	–	18
G	14	16.5	22	21	19	18	–

(a) Normally sweets are produced in the order A, B, C, D, E, F, G. The production line is then set up to start with flavour A the next day.

(i) Find the total time taken up by the changeovers.

(ii) Explain why this answer can be considered to be an upper bound for this travelling salesman problem.

(b) Use the nearest-neighbour algorithm, starting at A, to find a reduced time spent on changeovers.

(c) By initially ignoring sweet A, find a lower bound for the changeover times. [A]

11 A machine produces six different types of biscuit, A, B, C, D, E and F, one type at a time in any order. After producing each type of biscuit, the machine needs to be reset before producing the next type of biscuit. The times taken to reset the machine depend on the two types of biscuit involved and these times, in minutes, are given in the table. The machine is to produce each type of biscuit in turn before repeating the cycle. The machine can start the cycle with any type of biscuit.

From \ To	A	B	C	D	E	F
A	–	40	40	15	30	38
B	30	–	40	30	10	15
C	20	35	–	25	40	35
D	25	60	40	–	30	15
E	40	80	40	30	–	25
F	15	35	60	30	25	–

(a) **(i)** Use the nearest-neighbour algorithm, starting with A, to find an upper bound for the minimum total reset time in a complete cycle.

(ii) Explain why your answer to **(a)(i)** may be considered as being an upper bound for the minimum total reset time in a complete cycle.

(b) An algorithm for finding a lower bound is as follows.

Step 1 In each row find the minimum value and subtract this from all entries in that row and write down a matrix of adjusted times.

Step 2 For each column of adjusted times find the minimum value and subtract this from all entries in that column and hence write down a further matrix of times.

Step 3 To the sum of all the row minima you have found in step 1 add the sum of all the column minima you have found in step 2, to give a lower bound.

(i) Trace this algorithm for the biscuit machine, stating a lower bound for the total reset time.

(ii) Hence, or otherwise, find a cycle that will produce a total reset time of 135 minutes. [A]

12 A machine is used for producing sweets in six flavours. The machine produces one flavour of sweet at a time. It needs to be cleaned before changing flavours. The times taken to clean the machine depend on the two flavours involved and these times, in minutes, are given in the table below.

The machine is to be set to produce each flavour in sequence before repeating the cycle.

The machine can only start with raspberry (R) or strawberry (S).

To / From	**Blackcurrant** (B)	**Lime** (L)	**Orange** (O)	**Plum** (P)	**Raspberry** (R)	**Strawberry** (S)
Blackcurrant (B)	–	25	20	20	27	25
Lime (L)	15	–	10	11	15	10
Orange (O)	5	30	–	15	20	19
Plum (P)	20	16	15	–	25	10
Raspberry (R)	10	20	7	15	–	15
Strawberry (S)	15	25	19	10	20	–

(a) **(i)** Show that, using the nearest-neighbour algorithm starting with strawberry (S), the total cleaning time for one cycle is 85 minutes.

(ii) Use the same method starting with raspberry (R) to find the total cleaning time for one cycle, which is less than 85 minutes.

(b) Explain why each of your answers to **(a)** is an upper bound for the minimum total cleaning time for one cycle.

(c) Given that the machine produces S first followed by R, find an improved upper bound for the minimum total cleaning time for one cycle. [A]

13 Tony is going on a touring holiday in America. The following table shows the five cities that he is to visit. The figures represent the distances, in miles, between the cities.

	Los Angeles (LA)	**Las Vegas** (LV)	**Palm Springs** (PS)	**Santa Barbara** (SB)	**San Diego** (SD)
Los Angeles (LA)	–	190	210	90	185
Las Vegas (LV)	190	–	180	140	300
Palm Springs (PS)	210	180	–	230	150
Santa Barbara (SB)	90	140	230	–	250
San Diego (SD)	185	300	150	250	–

Tony is to start his tour at Los Angeles, visiting each city once before returning to Los Angeles.

(a) Use the nearest-neighbour algorithm, starting from Los Angeles, to find an upper bound for a tour of the five cities.

(b) By deleting Los Angeles, obtain a lower bound for a tour of the five cities.

(c) An optimal tour has length *T* miles. Using your answers to **(a)** and **(b)**, write down a statement about *T*. [A]

14 A researcher starts at the desk, D, of the British Library and wishes to consult books at locations E, F, G, H, I and J in the library. The time, in minutes, needed to walk between any two places is given in the table.

	D	**E**	**F**	**G**	**H**	**I**	**J**
D	–	7	6	9	7	10	8
E	7	–	11	6	12	10	7
F	6	11	–	10	11	11	12
G	9	6	10	–	13	11	8
H	7	12	11	13	–	8	12
I	10	10	11	11	8	–	9
J	8	7	12	8	12	9	–

(a) Use the nearest-neighbour algorithm, starting at D, to find one appropriate order in which the researcher might visit the six locations before returning to D. State the total time needed for walking around this particular route.

(b) Find a minimum connector of just the six locations E, F, G, H, I and J, and state its total length. Draw a tree representing your minimum connector.

(c) Explain why, in this case, the route you found in **(a)** is definitely the shortest possible. [A]

15 A delivery van based at A is required to deliver goods in the towns B, C, D, E shown in the diagram. The numbers on the edges are the distances in miles.

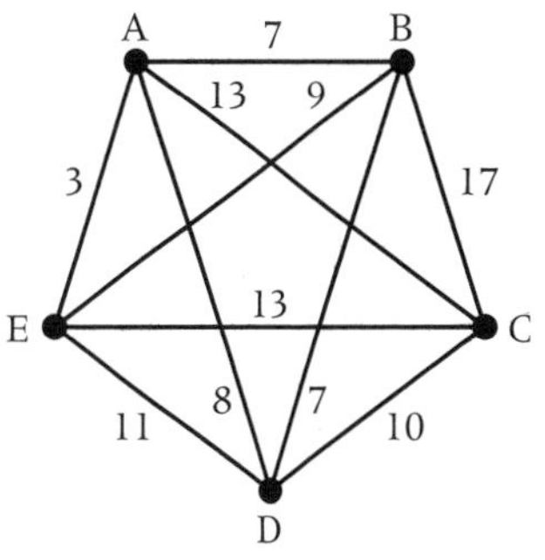

(a) Use the nearest-neighbour algorithm to find one possible route for the delivery van.

(b) Find a minimum connector of the network with A excluded.

(c) Use your answers to **(a)** and **(b)** to find upper and lower bounds for the distance the delivery van will have to travel. [A]

4

Key point summary

1 The **travelling salesman problem** is to find a tour visiting all vertices once before returning to the start vertex. A tour will always have the same number of edges as there are vertices in the network. *p65*

2 An **upper bound** is a value for the length of a tour that exists but may be improved upon. *p67*

3 A **lowest upper bound** is a value that cannot be improved upon. *p68*

4 The **nearest-neighbour algorithm** obtains an upper bound. *p68*

5 The best lower bound is the greatest lower bound. *p72*

6 The lower bound can be found using an algorithm. *p73*

7 The methods for finding upper and lower bounds only apply to **complete networks**. *p77*

Test yourself	What to review
1 List the number of possible tours around a complete network with five vertices labelled A, B, C, D and E.	*Section 4.1*
2 Apply the nearest-neighbour algorithm, starting from each vertex in turn, to find four upper bounds for the travelling salesman problem for the following network. State the best upper bound. 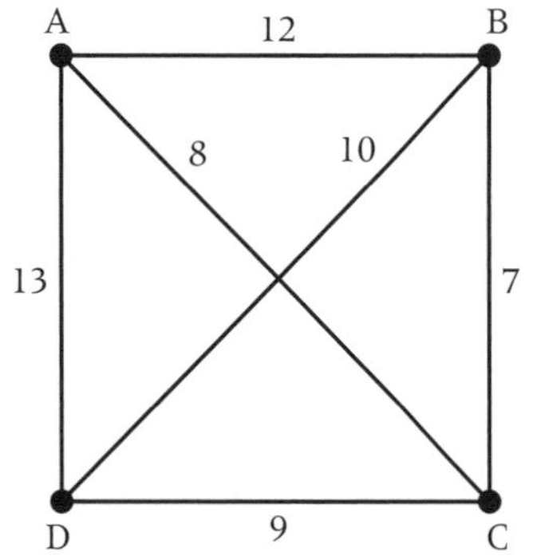	*Sections 4.2, 4.3*
3 By deleting each vertex in turn, find four lower bounds for the network in question 2. State the best lower bound.	*Section 4.5*
4 By considering your answers to questions 2 and 3, write down an interval of minimum width within which an optimal tour of the four vertices must lie.	*Sections 4.2, 4.3, 4.5*

Test yourself (continued)

What to review

5 The following network is incomplete.

Sections 4.2, 4.3, 4.5, 4.6

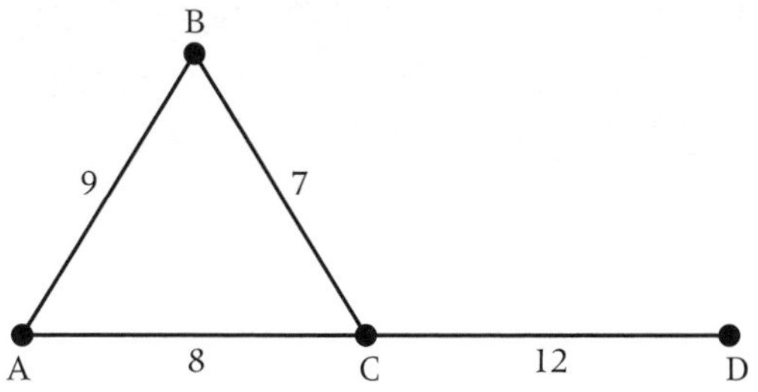

(a) List the minimum distances between pairs of vertices.

(b) Find the best upper and lower bound to your network.

(c) Write down the best upper and lower bound, referring to the original network.

Test yourself ANSWERS

1 ABCDEA ABCEDA ABDCEA ABDECA ABECDA ABEDCA
ACBDEA ACBEDA ACDBEA ACEBDA ADBCEA ADCBEA

2 ACBDA 38
BCADB 38
CBDAC 38
DCBAD 41
Best 38

3 Del A 36
Del B 34
Del C 37
Del D 34
Best 37

4 $37 \leqslant T \leqslant 38$

5 Upper bounds: ACBDA 54 BCADB 54 CBADC 48 DCBAD 48
Lower bounds: Del A 36 Del B 36 Del C 43 Del D 46
$44 \leqslant T \leqslant 48$, route CBACDC.

4

CHAPTER 5

Graph theory

Learning objectives

After studying this chapter, you should:

- know the meaning of the terms edge, vertex, tree, directed, trail, path and cycle
- be able to determine if a graph is Eulerian, semi-Eulerian or neither
- be familiar with different types of graphs: bipartite, complete, digraphs.

5.1 Introduction

Within this chapter we will look at adding some theory to the practical applications that we have already studied. It is essential that you have a clear understanding of the actual definitions and terminology used in graph theory.

5.2 Definitions

At GCSE we gained a basic understanding of what a graph actually is, for example $y = x^2 + 3x + 2$, but in this section the word graph is used in a more general sense.

A **graph** consists of a finite number of points connected by lines. Points are normally called **vertices** or **nodes**.

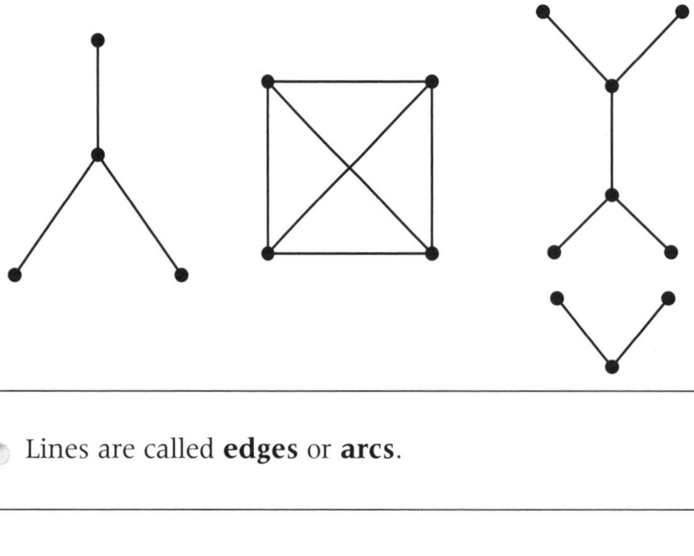

Lines are called **edges** or **arcs**.

If a graph has a number linked to each edge, then the graph is called a network (or weighted graph). The numbers on the edge may refer to time, distance or money.

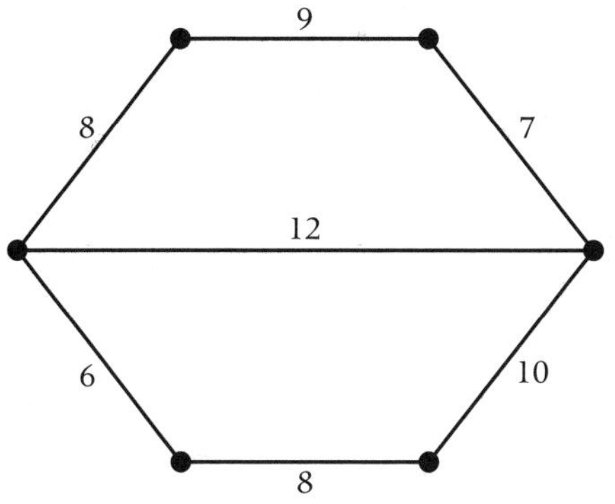

Two vertices are connected if there is an edge joining them.

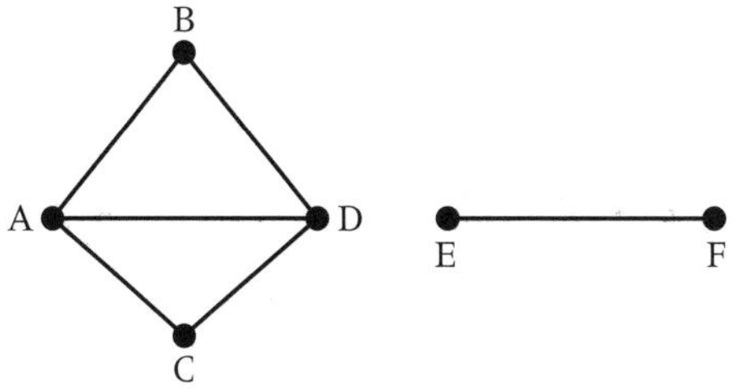

Edges AB, AC, AD, BD, CD and EF are connected, but AE, BE, CE, DE, AF, CF, DF and BC are not.

A graph is **connected** if all pairs of vertices are connected.

Note: a graph is said to be *fully connected* if every vertex is joined at least once to every other vertex.

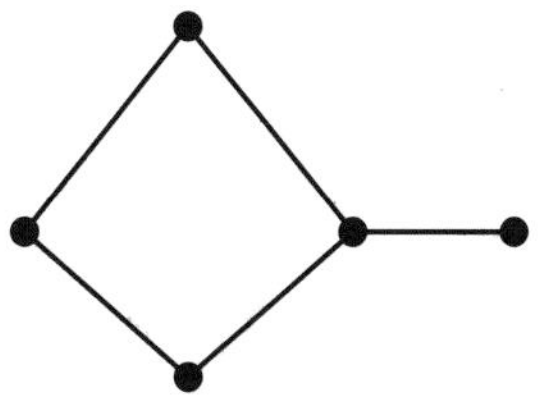

A connected graph

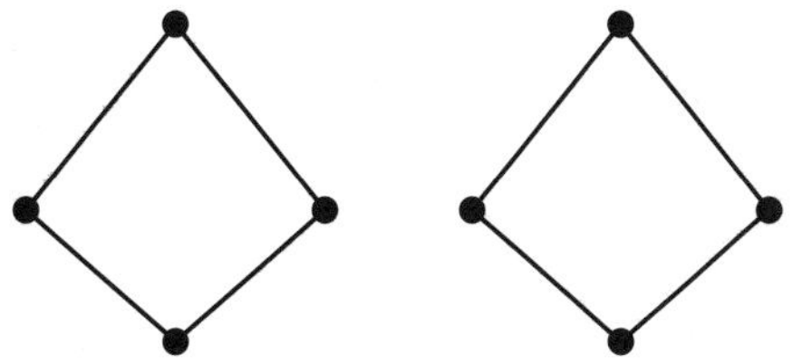

A disconnected graph

A simple graph is one in which there are no loops and at most one edge connects any pair of vertices.

A loop is an edge with the same vertex at each end.

5

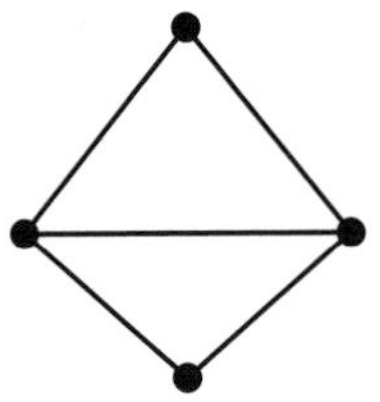

A simple graph

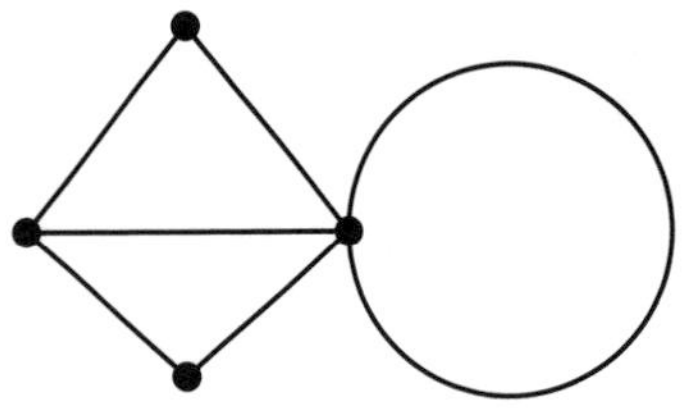

Not a simple graph

The **degree** (or order) of a vertex is the number of edges connected to the vertex.

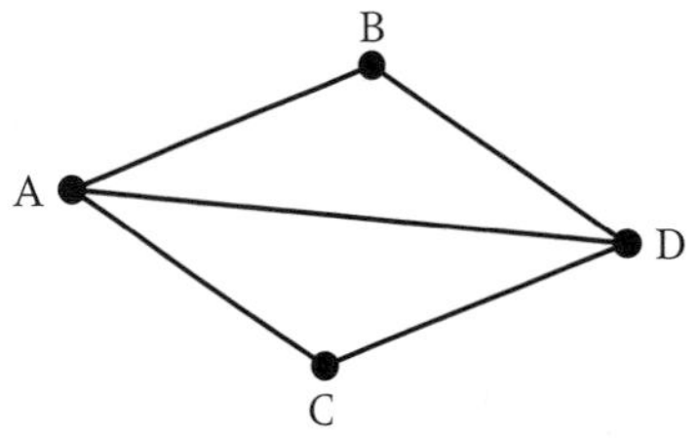

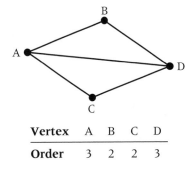

Vertex	A	B	C	D
Order	3	2	2	3

Worked example 5.1

A simple graph G has six vertices and their degrees are $2d$, $2d$, $2d + 1$, $2d + 1$, $2d + 1$ and $3d - 1$ where d is an integer.

(a) By considering the sum of all the degrees, show that d is even.

(b) Use the fact that the graph is simple to show that $d < 3$ and state the value of d.

(c) Draw a possible graph G.

Solution

(a) The sum of all the edges $= 13d + 2$, therefore d must be even.

(b) As the graph is simple, $3d - 1 < 5$, therefore $d < 2$.
As $d = 0$ is impossible, $d = 2$.

(c) A possible graph G is:

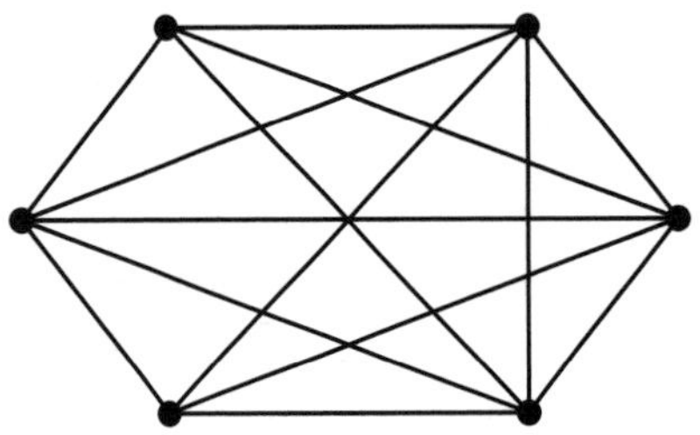

Sometimes the edges have a direction associated with them. For example a one-way road network system.

A graph that has directed edges is called a **directed graph** or **digraph**.

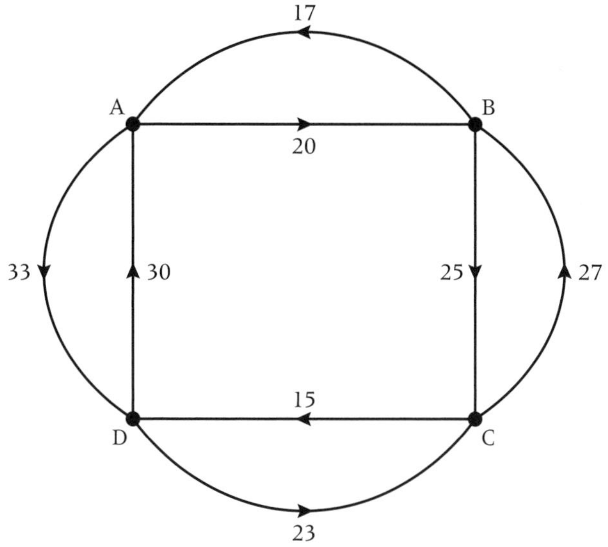

A **complete graph (K_n)** is a graph in which every vertex is connected by an edge to each of the other vertices.

For a complete graph with three vertices there must be $2 + 1 = 3$ edges.

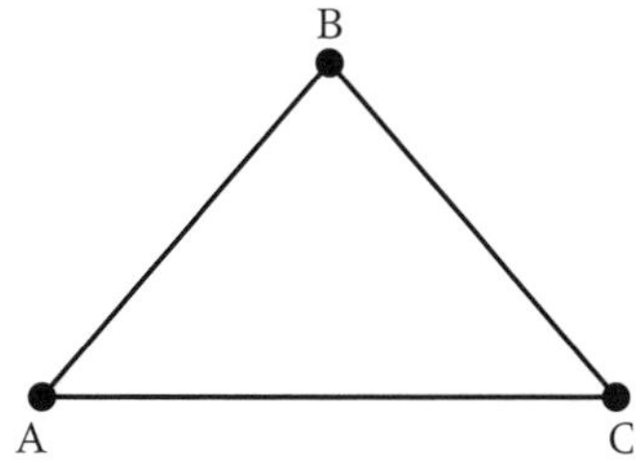

For a complete graph with four vertices the number of edges must be $3 + 2 + 1 = 6$ edges.

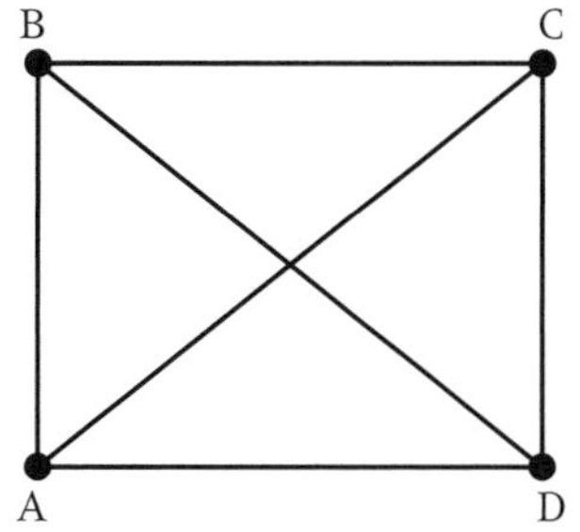

For a complete graph with five vertices there must be $4 + 3 + 2 + 1 = 10$ edges.

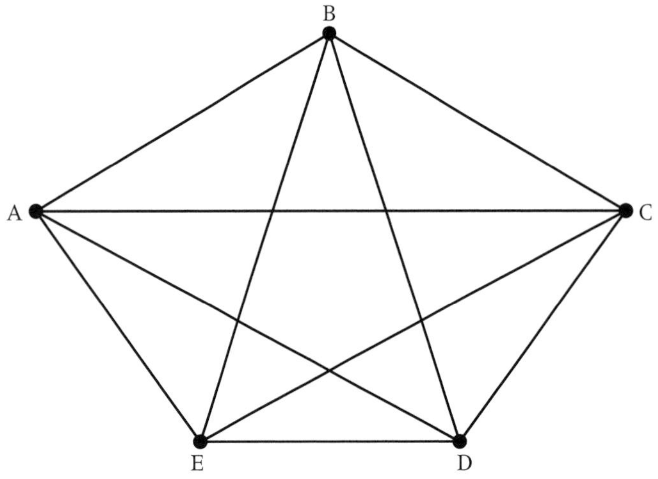

This information is summarised in the table below.

Number of vertices	Number of edges
3	3
4	6
5	10
6	15
n	$n(n-1)/2$

Worked example 5.2

A graph G has four vertices and edges of length 7, 8, 8 and 9 units.

(a) Explain why the graph G is not a complete graph.

(b) State the number of edges that must be added to G to make it a complete graph.

Solution

(a) As G has four vertices, there must be six edges in a complete graph.

(b) There are four edges given, so two extra edges must be added to make the graph complete.

A **bipartite graph** has two sets of vertices and the edges only connect vertices from one set to the other.

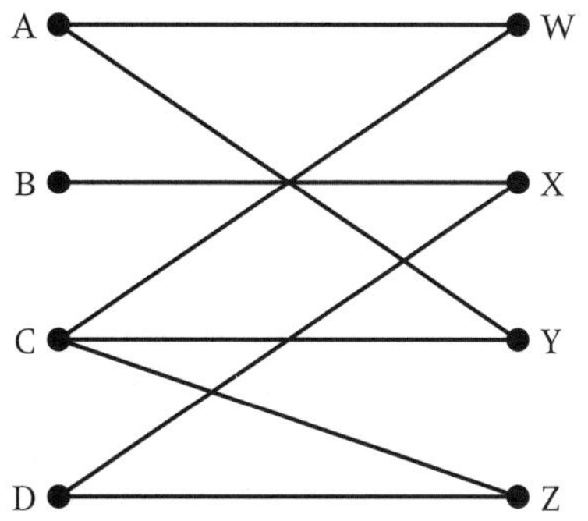

A **trail** is a sequence of edges of a graph such that the second vertex of each edge is the first vertex of the next edge, with no edge included more than once.

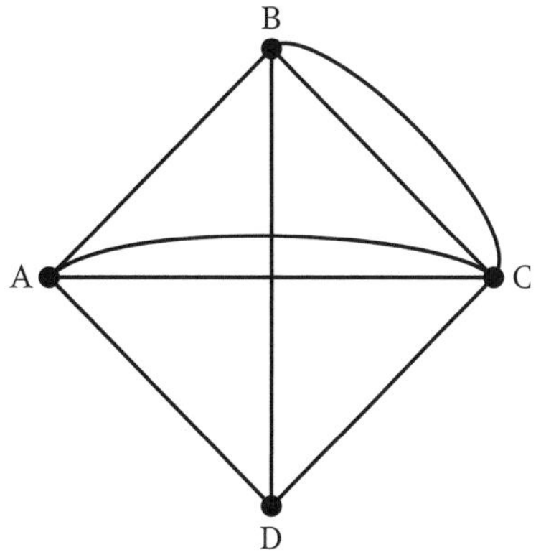

ABCADC is a trail.

A **path** is a trail such that no vertex is visited more than once (except that the first vertex maybe the last).

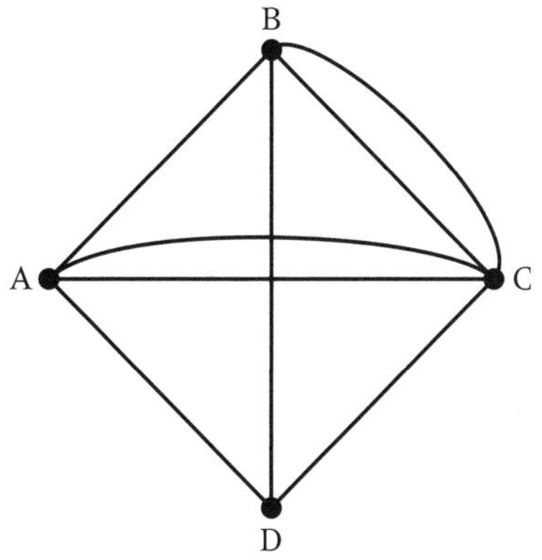

ABCD is a path.

A **cycle** is a closed path with at least one edge.

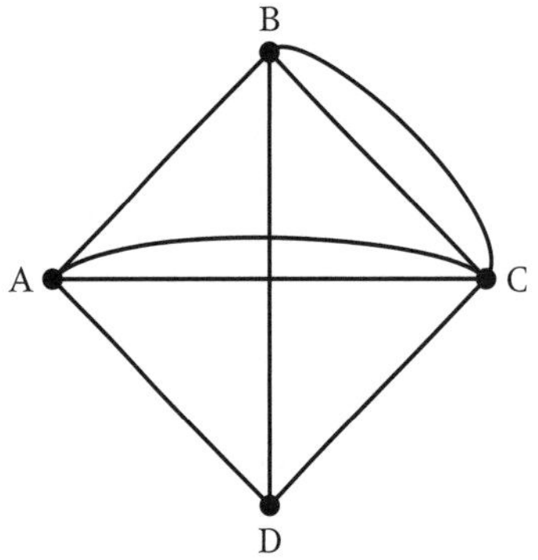

ABCA is a cycle.

A **Hamiltonian cycle** is a cycle that visits every vertex of a graph.

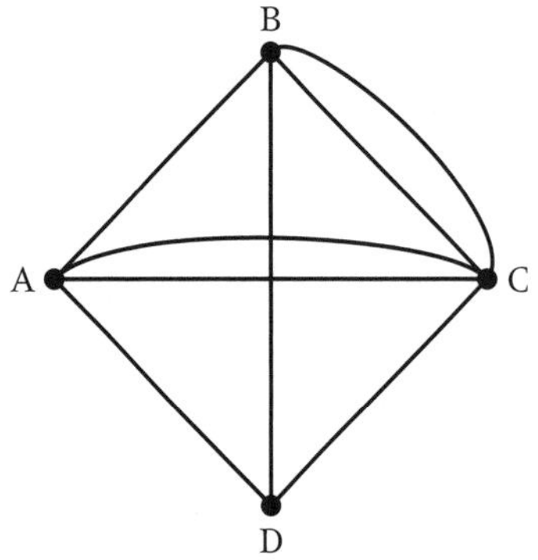

ABCDA is a Hamiltonian cycle.

Worked example 5.3

A complete graph G has four vertices labelled A, B, C and D.

(a) List the different Hamiltonian cycles of G.

(b) Find the number of different Hamiltonian cycles for a complete graph with

- **(i)** five vertices
- **(ii)** eight vertices.

(c) Write down the number of different Hamiltonian cycles for a complete digraph with

- **(i)** five vertices
- **(ii)** eight vertices.

Solution

(a) The different Hamiltonian cycles are:

ABCDA, ABDCA, ACBDA

(b) (i) There are $4!/2 = 12$ different Hamiltonian cycles.

(ii) There are $7!/2 = 2520$ different Hamiltonian cycles.

(c) (i) As we are now considering a digraph, then the cycle ABCDA is different to the cycle ADCBA, as the length AB is not necessarily the same as the length BA. There are $4! = 24$ different Hamiltonian cycles.

(ii) There are $7! = 5040$ different Hamiltonian cycles.

The cycles ACDBA, ADBCA, ADCBA might have been listed as well, but these cycles are the same as the ones already quoted: ABCDA = ADCBA

An **Eulerian trail** is a trail that uses all the edges of a graph. (If a graph possesses an Eulerian trail then the graph itself is called Eulerian. If the graph possesses a non-closed trail that uses all of its edges then the graph is called semi-Eulerian.)

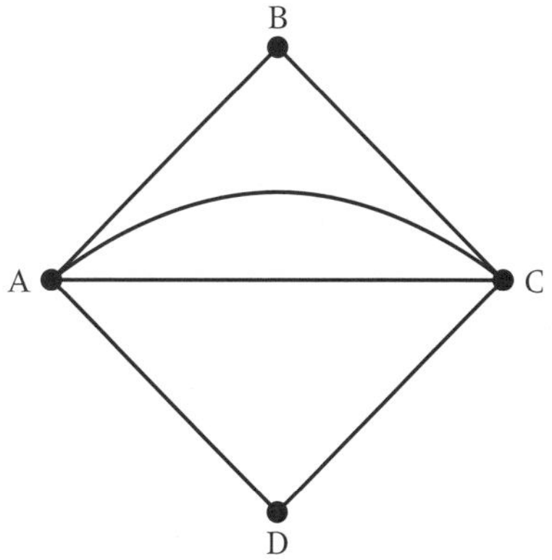

ABCDACA is an Eulerian trail.

Worked example 5.4

A simple graph, G, has five vertices and each of the vertices has the same degree d.

(a) State the possible values of d.

(b) If G is connected, what are the possible values of d?

(c) If G is Eulerian, what are the possible values of d?

Solution

(a) As there are five vertices, $d < 5$, therefore $d = 0, 1, 2, 3$ or 4.

(b) As the graph is connected, $d \neq 0, 1$.

(c) As the graph is Eulerian, $d \neq 3$, therefore $d = 2, 4$.

A **tree** is a connected graph with no cycles.

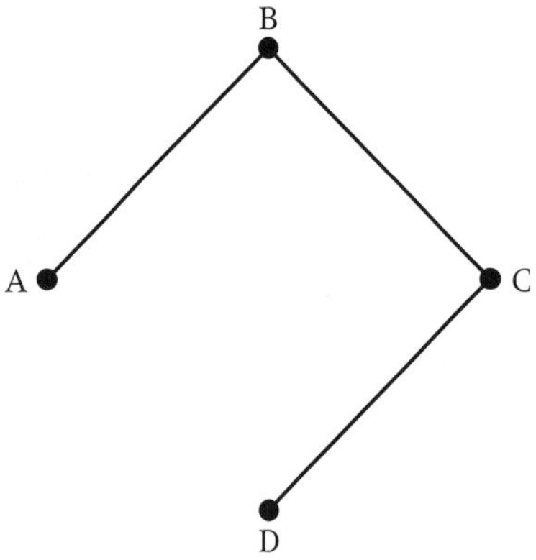

ABCD is a tree.

> Any tree that connects all the vertices of a graph is called a **spanning tree** for that graph. For a connected graph with n vertices each spanning tree has exactly $n - 1$ edges.

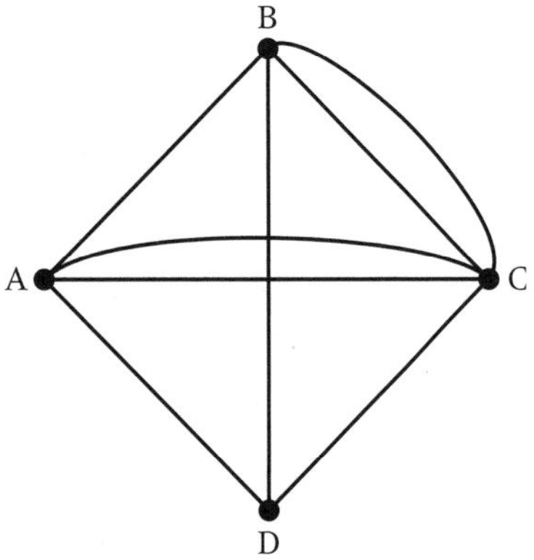

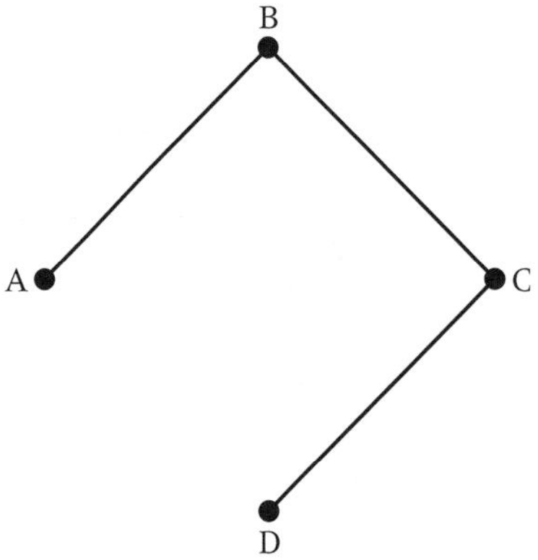

AB, BC, CD is a spanning tree of graph, G, on the left.

> A **minimum spanning tree** is a spanning tree of minimum weight for a network.

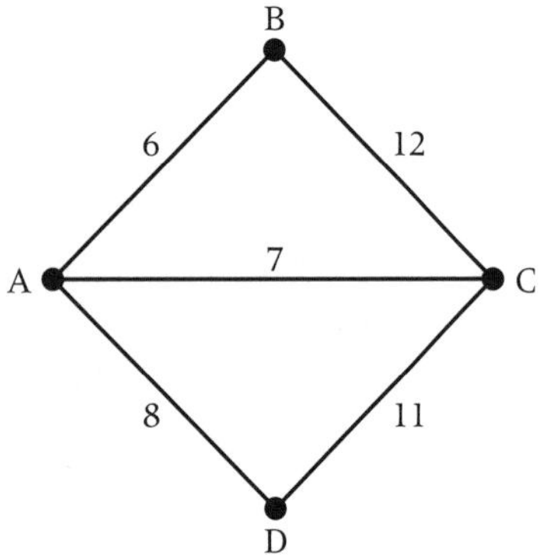

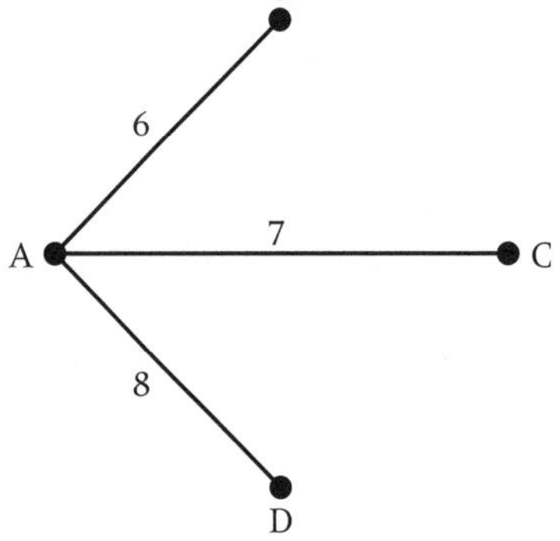

AB, AC, AD is a minimum spanning tree.

A graph may be represented by a matrix, which is called an adjacency matrix. Each row and column represent a vertex of a graph and the number in the matrix gives the number of edges joining the pair of vertices.

	A	**B**	**C**	**D**
A	0	1	2	1
B	1	0	2	0
C	2	2	0	1
D	1	1	1	0

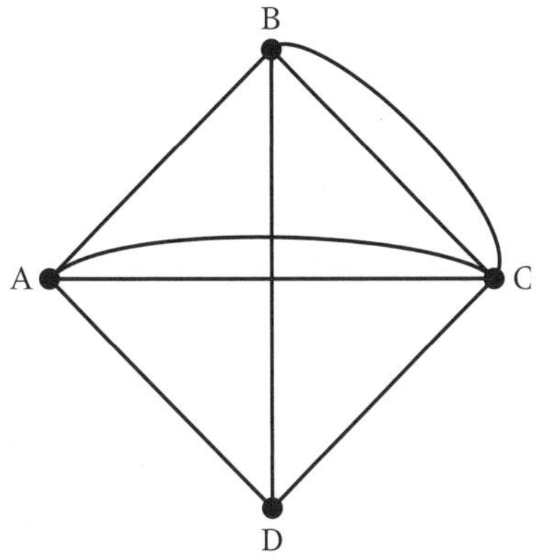

Worked example 5.5

A complete graph has an odd number of vertices and they all have degree *d*. Show that the graph is Eulerian.

Solution

As there is an odd number of vertices, each vertex must join to an even number of vertices, therefore *d* must be even and the graph is Eulerian.

Worked example 5.6

G is the graph illustrated, with six vertices and five edges:

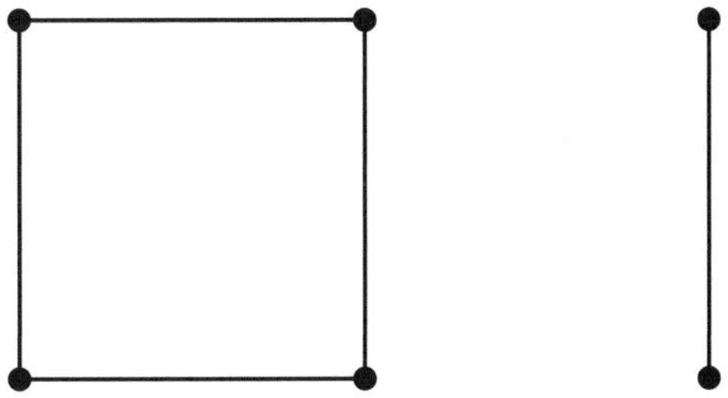

(a) What is the smallest number of edges that must be added to G in order to make a connected graph? Draw one such graph.

(b) What is the smallest number of edges that must be added to G in order to make a Hamiltonian graph? Draw one such graph.

(c) What is the smallest number of edges that must be added to G in order to make an Eulerian graph? Draw one such graph.

Solution

(a) One edge.

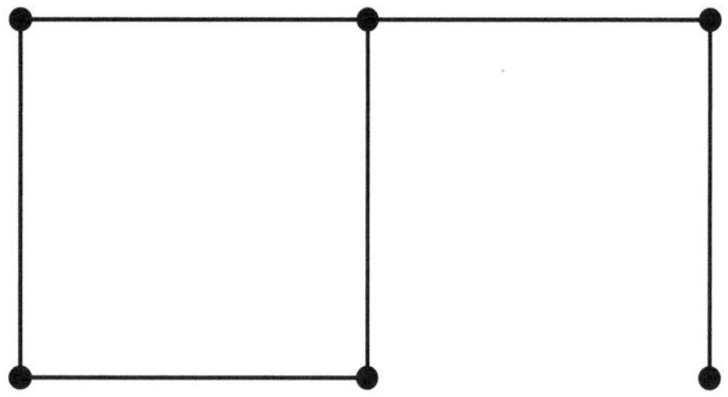

(b) Two edges.

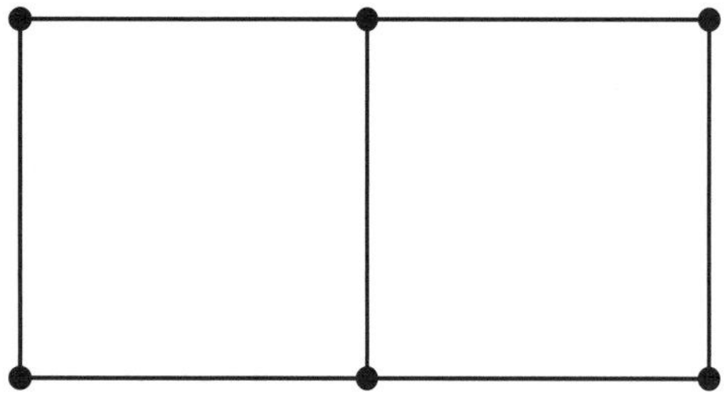

(c) Two edges.

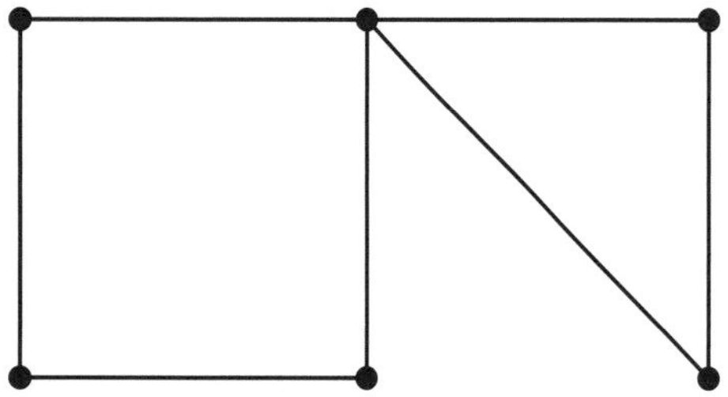

Worked example 5.7

A connected graph has five vertices and arc lengths of 6, 9, 9, 9, 10, 10, 12 and 15 units.

(a) State the minimum length of a minimum spanning tree for any such graph.

(b) State the minimum length of a Hamiltonian cycle for any such graph.

(c) State the minimum length of an Eulerian cycle for any such graph.

(d) In the case where the length of its minimum spanning tree is 34 units, draw a sketch of a possible graph.

Solution

(a) A minimum spanning tree has four edges, hence the minimum value is $6 + 9 + 9 + 9 = 33$ units.

(b) A Hamiltonian cycle has five edges, hence the minimum value is $6 + 9 + 9 + 9 + 10 = 43$ units.

(c) An Eulerian cycle must include all the edges at least once, hence the minimum value is 80 units.

(d) As the graph has five vertices, draw these first.

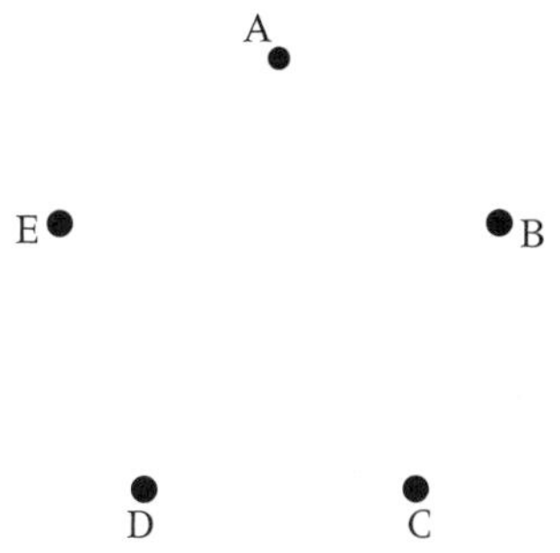

As the minimum spanning tree is 34, then it must include edges of length 6, 9, 9 and 10. Add edges corresponding to these values.

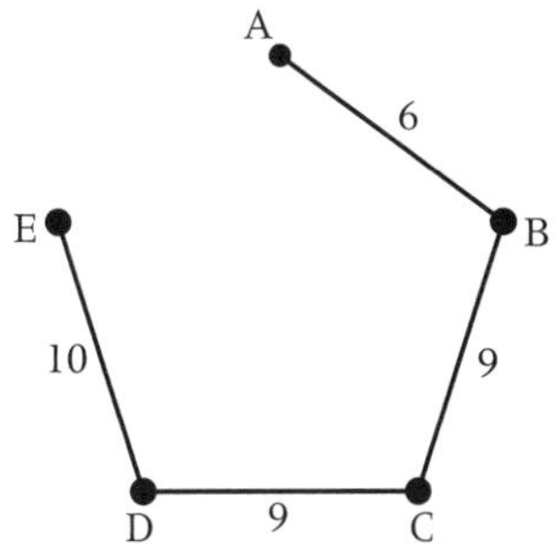

There are four edges to add to the graph. You must not add the edge AE of length 9, as this would alter the minimum spanning tree to length 33.
As there was no mention of the graph being simple then it is easy to add all the remaining edges to the edge AB, which will not affect the minimum spanning tree.

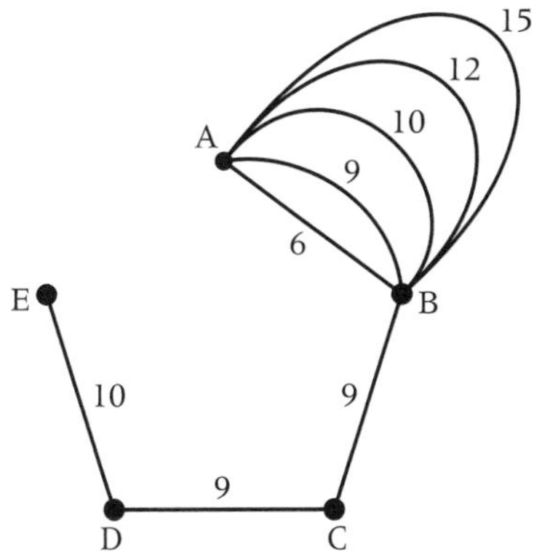

MIXED EXERCISE

1 **(a)** Explain what is meant by a Hamiltonian cycle.

(b) Determine how many different Hamiltonian cycles there are in a fully connected graph with four vertices.

(c) Write down the number of different Hamiltonian cycles there are in a fully connected graph with n vertices. [A]

2 **(a)** Draw a simple connected graph with 9 vertices all having degree 2.

(b) A simple connected graph has 9 vertices, all having the same degree d.

(i) State the possible values of d.

(ii) For each of these values of d, state the number of edges of the graph. [A]

3 A connected graph G has five vertices. The lengths of the edges are 7, 7, 7, 9, 11, 12, 13, 15 and 16 units, respectively.

(a) A graph is said to be *fully connected* if every vertex is joined at least once to every other vertex. Explain why graph G cannot be fully connected.

(b) A connected graph G is to be drawn with 5 vertices with lengths of edges as given above.

(i) Calculate the least possible length of a minimum spanning tree of graph G.

(ii) Explain why your answer to **(b)(i)** might not apply to graph G.

(iii) Sketch an example of graph G which has a minimum spanning tree of length 34 units. [A]

4 **(a)** Draw a graph to represent the following matrix.

(b) Explain from the matrix why your graph must be a directed graph.

From \ To	A	B	C	D
A	0	2	1	0
B	1	0	1	1
C	1	1	2	2
D	1	0	3	1

5 **(a)** **(i)** Draw the edges of a graph with the following vertex-to-vertex table.

(ii) Explain why this graph does **not** have an Eulerian cycle.

	A	B	C	D
A	0	1	2	1
B	1	0	2	1
C	2	2	0	1
D	1	1	1	0

(b) A vertex-to-vertex table is symmetrical about a leading diagonal consisting only of zeros. State the connection between the sum of all the numbers in the table and the number of edges in the corresponding graph.

(c) State the circumstances in which a graph of a vertex-to-vertex table is **not** symmetrical about its main diagonal. [A]

6 (a) A connected graph has four vertices. State the number of edges in the graph's minimum spanning tree.

(b) A graph has n vertices. The graph is complete, i.e. each vertex is joined to every other vertex by exactly one edge.

(i) State the number of edges in the graph's minimum spanning tree.

(ii) Determine the number of Hamiltonian cycles in the graph.

(c) A connected graph has four vertices and has arc lengths of 4, 4.5, 5, 6.5, 7, 8 and 9 units.

The length of its minimum spanning tree is 17 units. Draw a sketch to show a possible graph. [A]

7 A connected graph has five vertices and has arc lengths of 4, 7, 7, 7, 8, 8, 9 and 12 units.

(a) State the minimum length of a minimum spanning tree for any such graph.

(b) State the minimum length of a Hamiltonian cycle for any such graph.

(c) State the minimum length of an Eulerian cycle for any such graph.

(d) In the case when the length of its minimum spanning tree is 26 units, draw a sketch to show a possible graph. [A]

8 The government commissions the growing of a certain species of plant to assist in trials of a new drug. The plants must be grown in controlled conditions which necessitate regular watering using an automated sprinkler system. Each plant must have its own sprinkler outlet at its base.

The plants stand 1 m apart in rows. The rows are 2 m apart. The mains water supply is 5 m from the first plant, as shown in the following diagram.

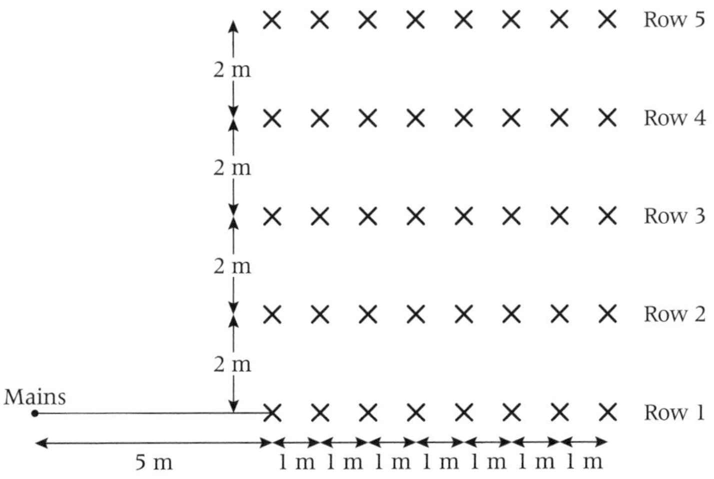

5

The minimum length of piping is to be used in the sprinkler system.

(a) Show that the minimum length of piping needed, if 365 plants are to be planted with 50 plants in each of seven rows and 15 plants in an eighth row, is 376 m.

(b) Find the minimum length of piping needed if 30 plants are to be planted.

(c) Find, in its simplest form, an expression for the minimum length of piping needed to plant m rows with each row containing exactly n plants.

(d) It is required to plant x plants, where x is a multiple of 60, in rows each containing 60 plants. Show that the minimum length of piping needed is $ax + b$, where a and b are numbers to be determined. [A]

9 **(a)** A simple connected graph has 7 vertices, all having the same degree d. Give the possible values of d, and for each value of d give the number of edges of the graph.

(b) Another simple connected graph has 8 vertices, all having the same degree d. Draw such a graph with $d = 3$, and give the other possible values of d. [A]

10 The travelling salesperson problem is to find the shortest Hamiltonian cycle in a network, i.e. the shortest path visiting each vertex **once and only once** and returning to the initial vertex.

(a) Regarding ABCDEA as different from AEDCBA, how many different Hamiltonian cycles are there in the graph K_5?

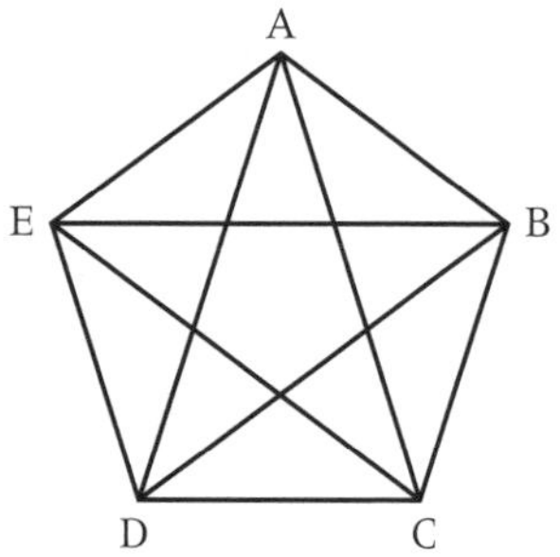

K_5

(b) How many Hamiltonian cycles are there in K_{50}?

A fast computer takes 1 s to find the lengths of 10 million such cycles. Approximately how many years will it take to find the lengths of all of them? [A]

11 **(a)** G_1 is a simple connected graph with 4 vertices.

(i) What is the least number of edges that G_1 could have?

(ii) What is the greatest number of edges that G_1 could have?

(b) G_2 is a simple connected graph with n vertices.

(i) What is the least number of edges that G_2 could have?

(ii) What is the greatest number of edges that G_2 could have? [A]

12 The plan below shows the rooms A–F of a museum, the gaps in the walls representing doors between the rooms.

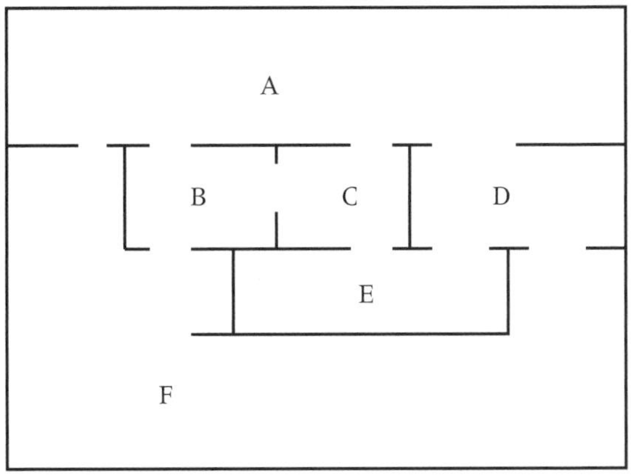

(a) Draw a graph with six vertices labelled A, B, C, D, E and F in which two vertices are joined by an edge if there is a door directly from one of the rooms into the other. So, for example, there will be an edge DE but there will not be an edge AE.

(b) State whether your graph is Eulerian, semi-Eulerian or neither. Give a reason for your answer.

(c) Give an example of a Hamiltonian cycle in your graph.

(d) If you had to design the layout of a museum would you prefer its graph to be Eulerian or Hamiltonian? Give a reason for your answer. [A]

13 The graph G is illustrated.

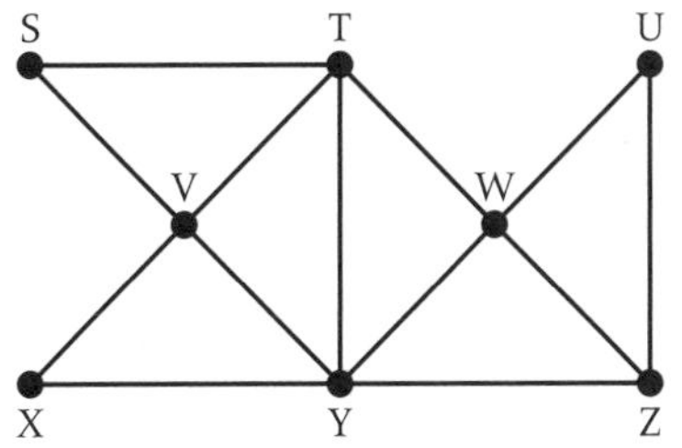

(a) (i) Explain how you know that graph G is semi-Eulerian.

(ii) Give an example of an Eulerian trail of G which begins ZYTW …

(b) Give a Hamiltonian cycle of G which begins ZU …

(c) (i) State which single edge should be removed from G in order to obtain an Eulerian graph.

(ii) Explain briefly how you know that the Eulerian graph obtained in **(c)(i)** is not Hamiltonian. [A]

14

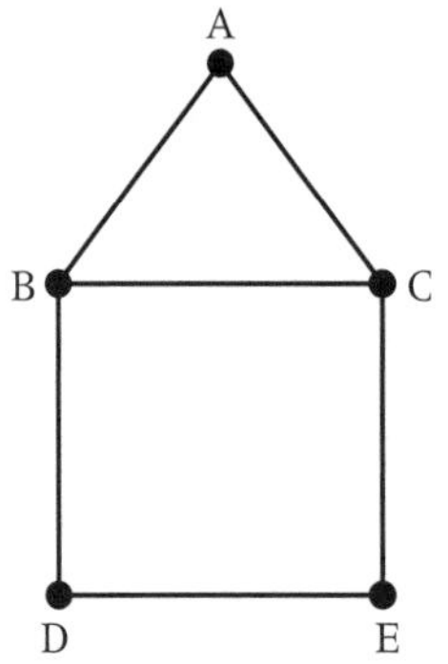

Graph G

Two of graph G's spanning trees (or minimum connectors, with all the edges of equal weight) T_1 and T_2 are shown below. In each of these trees the degree of the vertex A equals the degree of the vertex D.

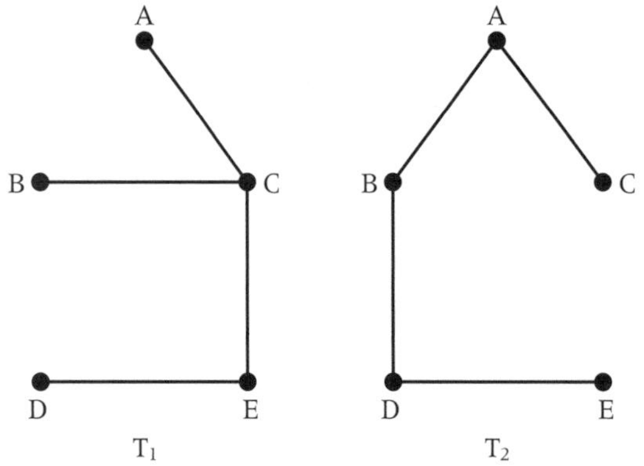

Draw **three** other spanning trees of G in which the degree of vertex A equals the degree of vertex D. [A]

Key point summary

1	A **graph** consists of **vertices** and **edges**. The degree of a vertex is the number of edges connected to a vertex.	*p90*
2	A network has numbers linked to each edge.	*p91*
3	A graph is connected if all pairs of vertices are connected. It is simple if there are no loops.	*p91*
4	The **degree** is the number of edges connected to the vertex.	*p92*
5	A **directed graph** has directed edges.	*p93*
6	A **complete graph** $\mathbf{K}_n$ consists of n vertices with each vertex joined to each other vertex once.	*p93*

Key point summary continued

7	A **bipartite graph** has two sets of vertices and the edges only connect vertices from one set to the other.	*p94*
8	A **trail** is a sequence of edges that only includes edges once.	*p95*
9	A **path** is a trail that only uses a vertex once.	*p95*
10	A closed path is a **cycle**.	*p95*
11	A cycle that visits all vertices is a **Hamiltonian cycle**.	*p96*
12	An **Eulerian trail** is a trail using all the edges of a graph. For an Eulerian trail to exist the degree of each vertex must be even.	*p97*
13	A **tree** is a connected graph with no cycles.	*p97*
14	A **spanning tree** is a simple connected graph with one fewer edge than vertices.	*p98*
15	A **minimum spanning tree** has minimum weight.	*p98*
16	A graph can be represented by an adjacency matrix.	*p99*

Test yourself	What to review
1 A simple graph has four vertices with degrees 2, 2, 2 and d. What are the possible values of d? Draw one example for each possible d.	*Section 5.2*
2 Draw the graph K_4.	*Section 5.2*
3 Write down an Eulerian trail of the graph K_5 (with vertices A–E). Is the graph K_6 Eulerian?	*Section 5.2*
4 Draw a graph G with four vertices and edges of length 2, 3, 4, 5 and 6 such that the minimum spanning tree is: **(a)** 9, **(b)** 10, **(c)** 13.	*Section 5.2*
5 How many different Hamiltonian cycles are there in K_5? How many Hamiltonian cycles would there be if the graph was a digraph?	*Section 5.2*

Test yourself ANSWERS

1 $d = 0, 2$

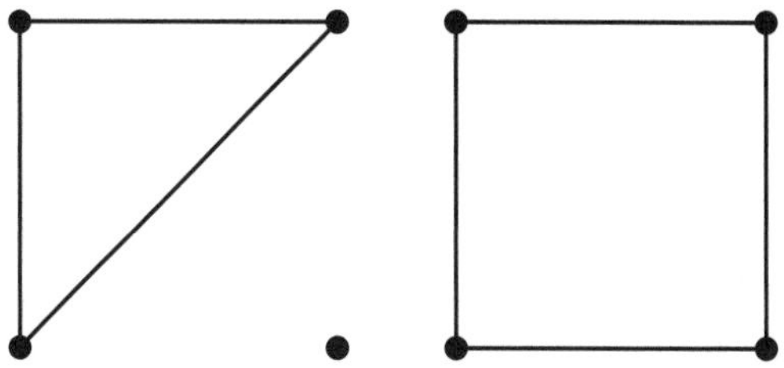

2

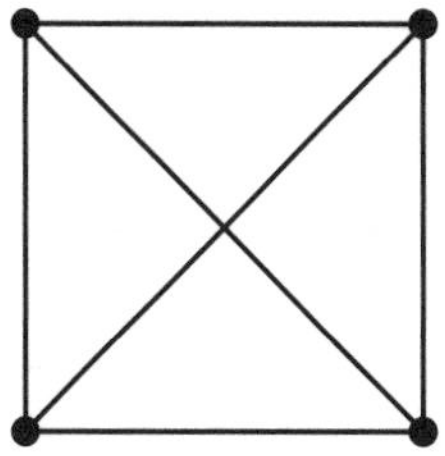

3 ABCDEADBECA
K is not Eulerian, as vertices are of odd order.

4 (a)

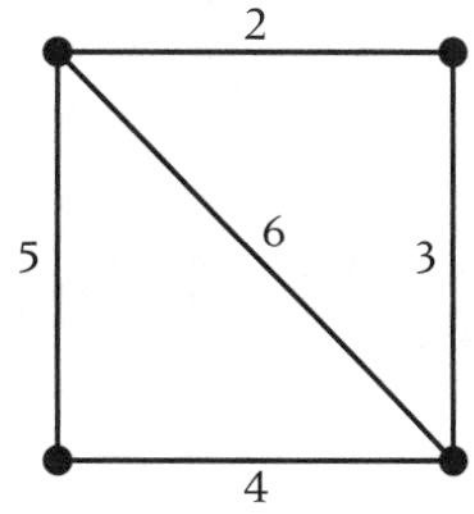

(b)

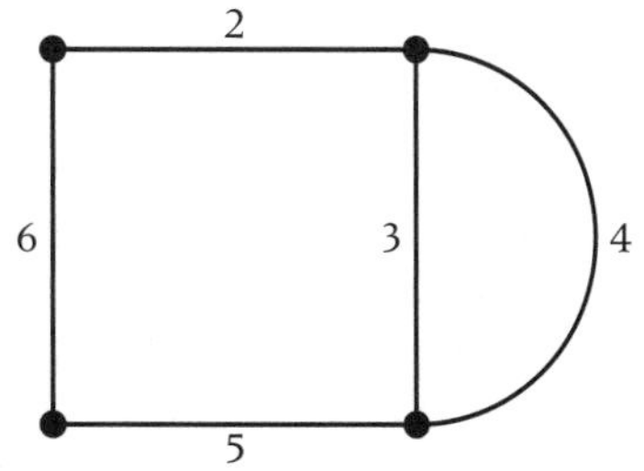

(c)

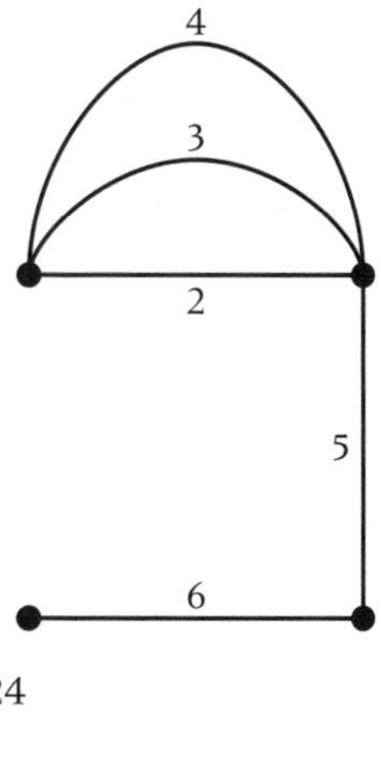

5 12, 24

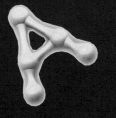

CHAPTER 6

Matchings

Learning objectives

After studying this chapter you should be able to:

- understand the relevance and use of a bipartite graph
- express a matching problem as a bipartite graph
- apply an alternating path to find a maximal match.

6.1 Introduction

Consider the following problem. Four boys, Alan (A), Bob (B), Colin (C) and Dave (D), are going to be paired with four girls, Serena (S), Rose (R), Tanya (T) and Viv (V). The girls have decided that Serena will only be paired with Bob, Rose will be paired with anyone other than Dave, Tanya will only be paired with Bob or Colin, and Viv insists on being paired with Alan. How can a match be arranged that will have each boy paired with a different girl?

To solve this problem, you could list all possible pairings and decide which will satisfy the problem. This is acceptable as we only need to find four matchings. However, using this method to solve a similar problem with 30 boys and 30 girls would be impractical.

We could consider the initial solution as an upper bound: we know that the initial situation is possible but we may be able to improve this and obtain a better match.

We can solve this type of problem using a **bipartite graph**, where from an initial attempt at a solution we will apply an algorithm to try to improve on this solution.

A bipartite graph G consists of two sets of vertices X and Y. The edges in G join vertices in X to vertices in Y. A bipartite graph for the example of the boys pairing with girls is shown below.

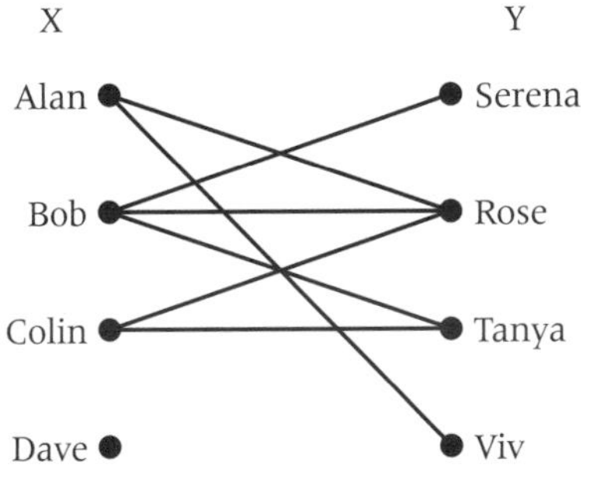

The set X contains the boys Alan, Bob, Colin and Dave, and the set Y contains the girls Serena, Rose, Tanya and Viv. A link between one of the boys in X and one of the girls in Y indicates that that particular boy can be paired with a particular girl. For example, Alan can be paired with Rose or Viv, but not with Serena or Tanya. This information may be represented by an adjacency matrix. The vertices on the left-hand side are written down the side and the vertices on the right-hand side are written along the top. Edges that link the vertices are represented by a 1, and where there is no connection by a 0.
For example, the bipartite graph shown above is represented by the matrix below.

	S	**R**	**T**	**V**
A	0	1	0	1
B	1	1	1	0
C	0	1	1	0
D	0	0	0	0

Worked example 6.1

Four pupils, A, B, C and D, are sitting around a table and are each going to colour in a picture. The colours of the pencils that they have available are red, white, green and pink. The pencils that each pupil wishes to use are listed below.

Ahmed (A)	red (R) and white (W)
Barry (B)	white (W) and green (G)
Colin (C)	green (G) and pink (P)
Danny (D)	white (W) and green (G)

Show this information on a bipartite graph.

Solution

This information can be modelled by using the bipartite graph below.

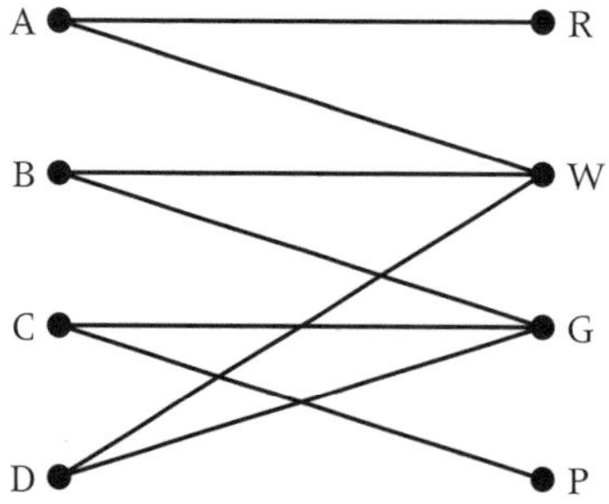

If each pupil is to be given one colour, it is clear that Ahmed has to be given red and Colin has to be given pink. White and green colours could be given to either Barry or Danny.

EXERCISE 6A

1 Four secretaries in an office were asked which five jobs they would like to do. Their replies are summarised below.

1	A, C, E
2	B, C, D
3	A, E
4	C, D, E

Represent this information in a bipartite graph.

2 A bridge team consists of eight players: four men and four women. The four men's names are Alan (A), Barveen (B), Chris (C) and Daniel (D). The four girls are called Mia (M), Nuala (N), Pat (P) and Ruth (R). Each man is going to be paired with a woman. Alan can be paired with anybody but Ruth. Barveen will play with either Pat or Ruth. Chris can play with Mia or Ruth, and Daniel can be paired with anyone except Mia.

Represent this information by a bipartite graph showing which men may be paired with which women.

3 An outward bound centre employs five instructors: Amy (A), Brendan (B), Craig (C), Dyllis (D) and Eric (E). The activities that this centre offers are rock climbing (R), kayaking (K), orienteering (O), problem-solving (P) and team-building (T).

The table below lists the staff and the activities they are able to teach.

A	rock climbing (R) and orienteering (O)
B	rock climbing (R) and team-building (T)
C	rock climbing (R), kayaking (K) and team-building (T)
D	rock climbing (R) and problem-solving (P)
E	rock climbing (R), team-building (T) and orienteering (O)

Represent this information in a bipartite graph.

4 A graph G is represented by the adjacency matrix below.

	V	**W**	**X**	**Y**	**Z**
A	1	0	1	0	1
B	0	1	1	1	0
C	1	1	0	1	0
D	1	0	1	0	1
E	0	0	1	1	1

Represent this information as a bipartite graph.

6.2 Matchings

We have modelled various situations using a bipartite graph. In the introduction to the chapter the following problem was introduced. Four boys, Alan (A), Bob (B), Colin (C) and Dave (D), are going to be paired with four girls, Serena (S), Rose (R), Tanya (T) and Viv (V). The girls have decided that Serena will only be paired with Bob, Rose will be paired with anyone other than Dave, Tanya will only be paired with Bob or Colin, and Viv insists on being paired with Alan.
How can a match be arranged that will have each boy paired with a different girl?

Pairing vertices of one of the sets, the boys, to vertices of the other sets, the girls, in such a way that no two edges have a common vertex is called a **matching**.

A **maximum matching** is any matching that contains the largest possible number of edges.

A **complete matching** is a matching that contains the same number of edges as there are vertices on either of the two sets.

Although from any set of information a maximum matching is always possible, a complete matching is not always possible.

Worked example 6.2

Three boys want to go to a dance with three girls. Alan will go to the dance with Rebecca, Sarah or Teresa. Bill will only go to the dance with Teresa. Colin will only go to the dance with Teresa. Is a complete match possible?

Solution

If we draw a bipartite graph of this situation it will become obvious that Bill and Colin will only go to the dance with Teresa and hence a complete matching is impossible. The maximum matching in this case will encompass two edges that include Alan and either Bill or Colin. This is illustrated in the bipartite graph below.

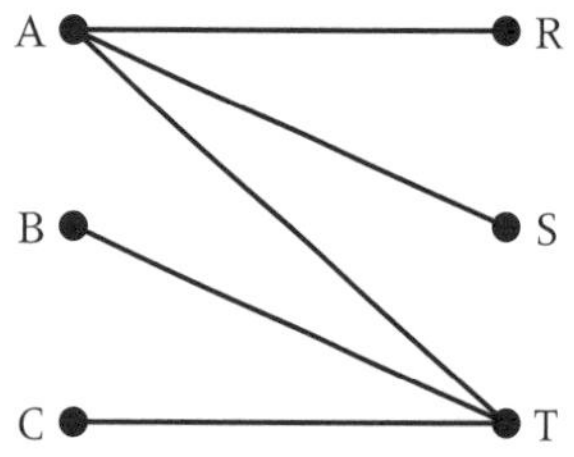

EXERCISE 6B

1 Explain why a complete matching is impossible in the bipartite graph below.

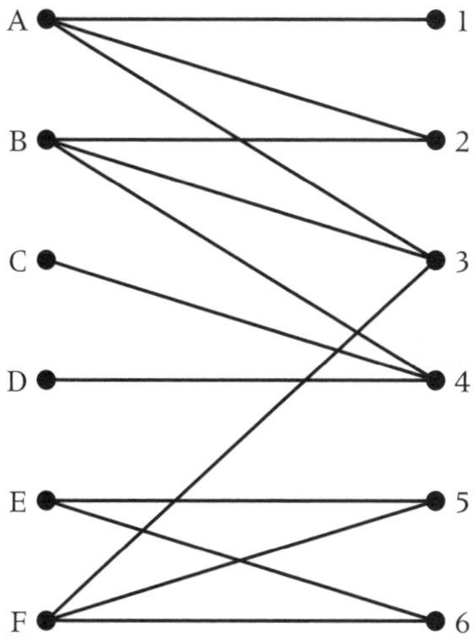

2 Find the number of distinct complete matchings that are possible using the bipartite graph below.

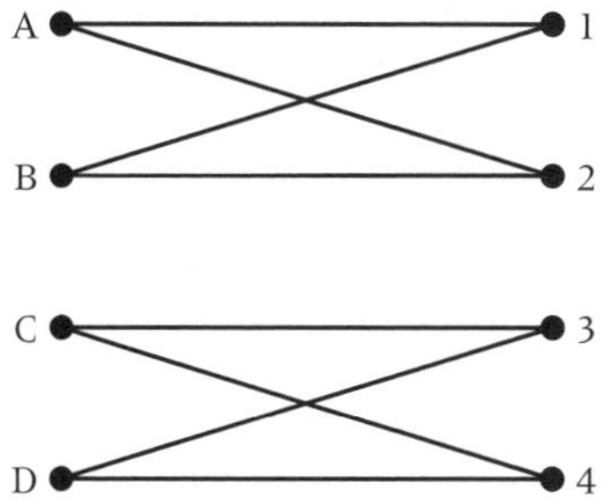

3 Show that there is only one complete matching for the bipartite graph below.

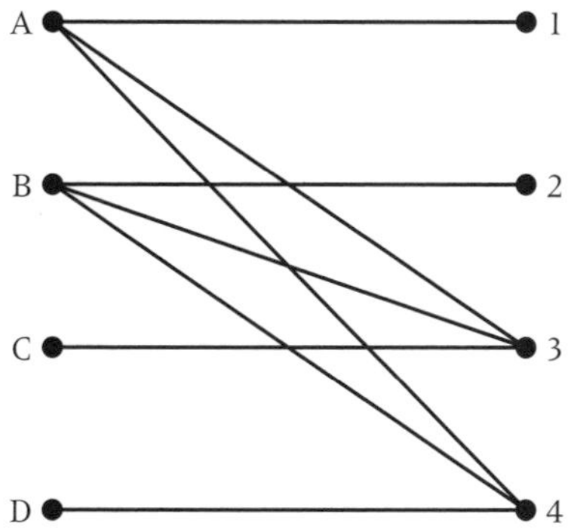

6.3 Improving a matching to obtain a maximal matching

For many bipartite graphs we will normally be presented with the problem of finding the largest possible set of edges that pair two sets of vertices, given that no vertex is to be joined to more than one vertex in the other set. We start from an initial situation and try to improve on this to obtain a maximum match.

> The principle is to start with a vertex in one set that is not in the initial match and alternate between the two sets until we finish at a vertex in the other set that was not in the initial match. This is called an **alternating path**.

> An algorithm for completing this is:
>
> **Step 1** From your initial matching, find a vertex on the left-hand subset not in the initial match and connect this vertex to a vertex on the right-hand set.
>
> **Step 2** If the right-hand vertex is not in the initial matching then add this to the initial matching and repeat step 1. If the right-hand vertex is in the initial matching go to step 3.
>
> **Step 3** Add the new edge to the matching and remove from the initial matching the edge linking this vertex to the left-hand side. Repeat step 1 using the vertex that has just been removed from the matching.
>
> **Step 4** Continue in this way until you have a complete matching or there is no further improvement that can be made.

Consider the problem at the start of the chapter. Four boys, Alan (A), Bob (B), Colin (C) and Dave (D), are going to be paired with four girls, Serena (S), Rose (R), Tanya (T) and Viv (V). The girls have decided that Serena will only be paired with Bob, Rose will be paired with anyone other than Dave, Tanya will only be paired with Bob or Colin, and Viv insists on being paired with Alan. Given that initially Rose is paired with Bob and Tanya is paired with Colin, can a match be arranged that will have each boy paired with a different girl?

Draw a bipartite graph to represent the information and then apply the algorithm to the problem.

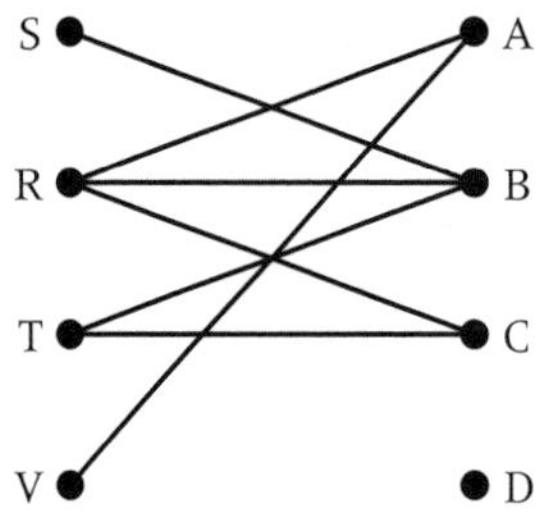

The initial match is shown on the diagram below.

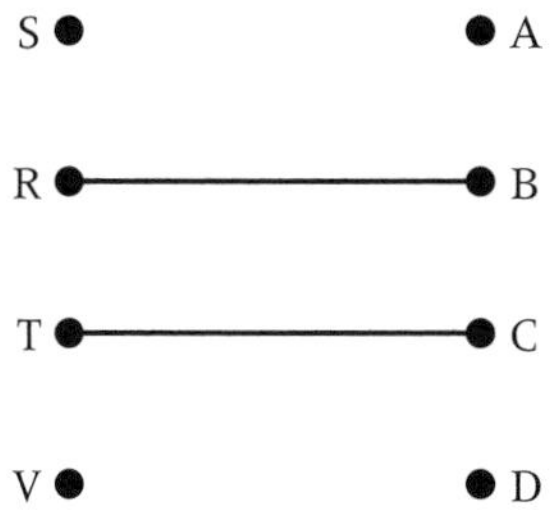

Step 1 Serena (S) is not in the initial match. Connect Serena to Bob (B).

Step 2 As the right-hand vertex is in the initial match, go to step 3.

Step 3 Add Serena to Bob to the match. The edge Bob to Rose (R) is now removed from the match.

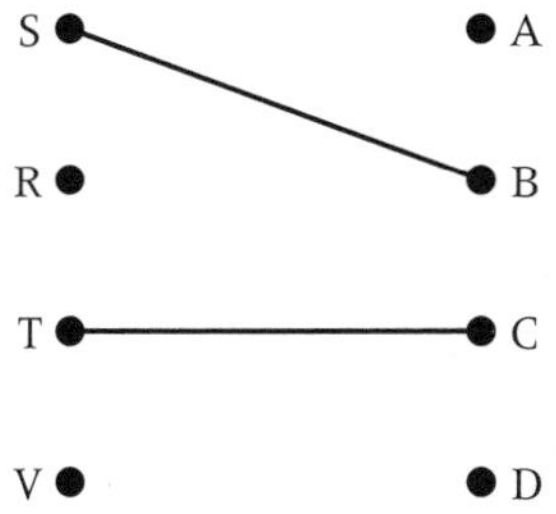

Step 1 Rose is now not in the match. Connect Rose to Alan (A).

Step 2 The right-hand vertex is not in the match, hence add Rose to Alan.

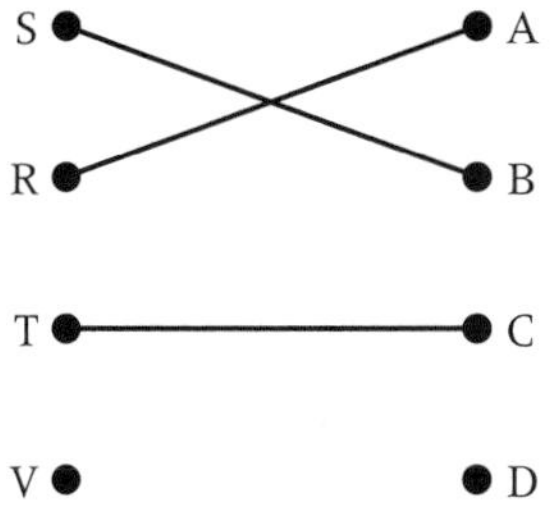

The only boy not in the match is Dave. The bipartite graph indicates that Dave cannot be matched with a girl, consequently a **maximum match** has been obtained. As there is a vertex not included in the final match a **complete match** has not been found.

Worked example 6.3

There are five workers, A, B, C, D and E, and five tasks to which they are to be assigned, 1, 2, 3, 4 and 5. The table below lists the tasks to which they can be assigned.

A	1, 2
B	1, 4, 5
C	2, 3
D	2, 5
E	3, 5

If the initial match pairs together A – 2, B – 1, C – 3 and D – 5, find a complete match of workers to tasks.

Solution

We are trying to find a patch from an unconnected vertex on the left-hand set to an unconnected vertex on the right-hand set. We can represent the original information using a bipartite graph. Alongside this graph we draw another bipartite graph that shows the initial match.

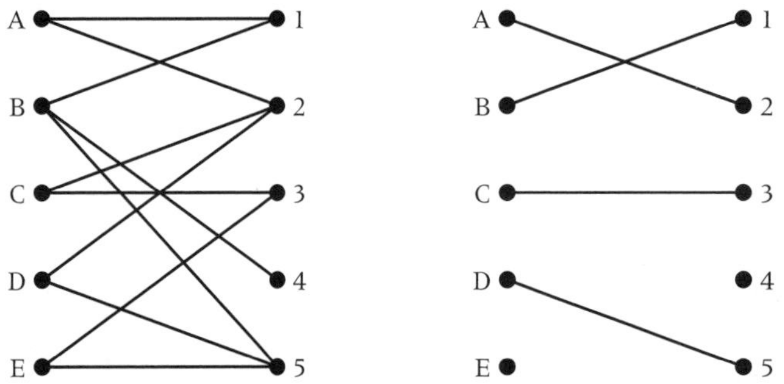

The graph on the left indicates possible edges that might be included in our match, and the graph on the right indicates our starting solution.
Using the algorithm we need to find a path to connect the unconnected vertex E in set X to the unconnected vertex 4 in set Y. From E, connect to either 3 or 5.

(a) If initially we connect E to 3, then 3 cannot be connected to C, which we now connect to 2, which we disconnect from A, which we now connect to 1, which we disconnect from B, which we now connect to 4.
Writing this as a path gives E – 3 ≠ C – 2 ≠ A – 1 ≠ B – 4.
Connections from left to right are in the new match, but connections from right to left are removed from the match.
D was originally connected to 5 and that match is unaltered.
This gives our final match as (A – 1), (B – 4), (C – 2), (D – 5) and (E – 3).

(b) If initially we connect E to 5, then our path becomes E – 5 ≠ D – 2 ≠ A – 1 ≠ B – 4.
Originally C was connected to 3 and this match is unaltered.
This gives our final match as (A – 1), (B – 4), (C – 3), (D – 2) and (E – 5).

Notice that our two alternating paths lead to different, but equally correct, solutions.

Worked example 6.4

A new set of four mathematics text books are to be written. The four books are on mechanics (M), pure (P), networks (N) and statistics (S). The four authors are Anthony (A), David (D), Edward (E) and John (J).
The books that the authors are prepared to write are given in the table below.

Anthony (A)	mechanics (M), pure (P)
David (D)	networks (N), pure (P)
Edward (E)	mechanics (M), pure (P)
John (J)	networks (N), statistics (S)

Initially Anthony is matched to mechanics, David to pure and John to statistics.
Represent the information using a bipartite graph and from the initial match use an algorithm to find a complete match of authors to books.

Solution

The bipartite graph and the initial match are as shown.

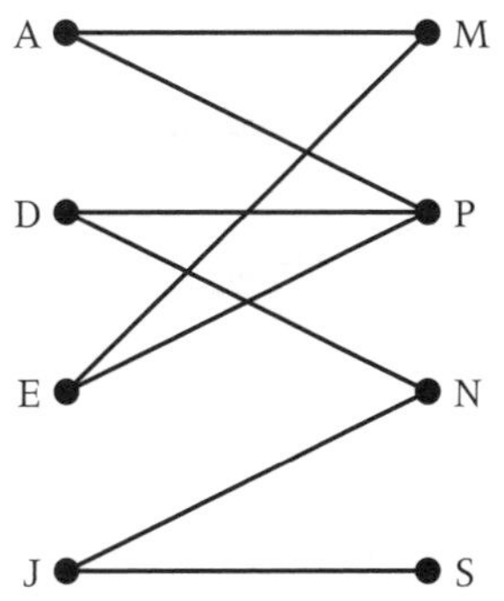

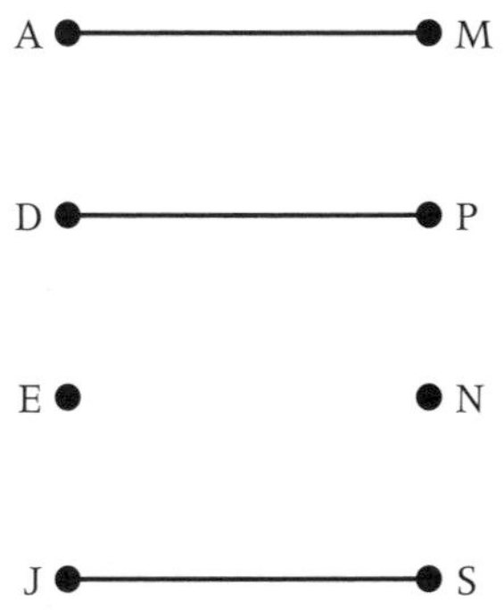

The unconnected vertex is E and this gives two potential alternating paths:

E – P ≠ D – N
E – M ≠ A – P ≠ D – N

These alternating paths lead to the following complete matches: (A – M), (D – N), (E – P), (J – S) and (A – P), (D – N), (E – M) and (J – S).

Both of these solutions are equally valid.

EXERCISE 6C

1 The bipartite graph below shows how two sets of vertices may be connected.

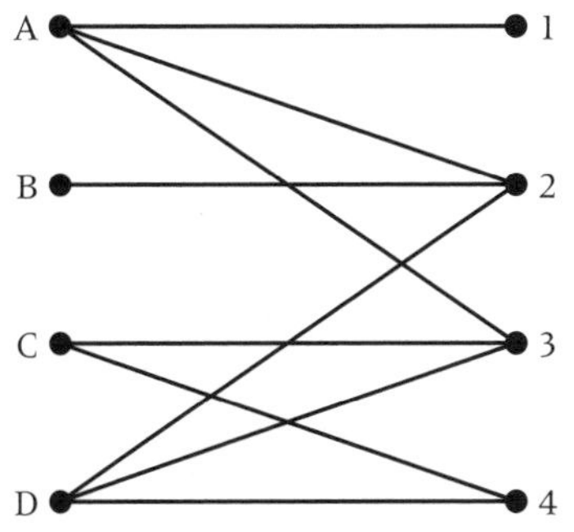

The initial match is A – 1, C – 4 and D – 2.
From this initial match, apply an algorithm to find a complete match.

2 The adjacency matrix shown below represents an incomplete matching. (A number written in bold indicates that the edge is in the initial match.)

	P	Q	R	S	T
A	1	**1**	1	0	0
B	0	1	0	0	**1**
C	**1**	0	1	0	1
D	0	0	**1**	1	0
E	0	0	1	0	1

From this initial match, use an alternating path to find a complete match.

3 The bipartite graph below indicates how two sets of vertices may be connected.

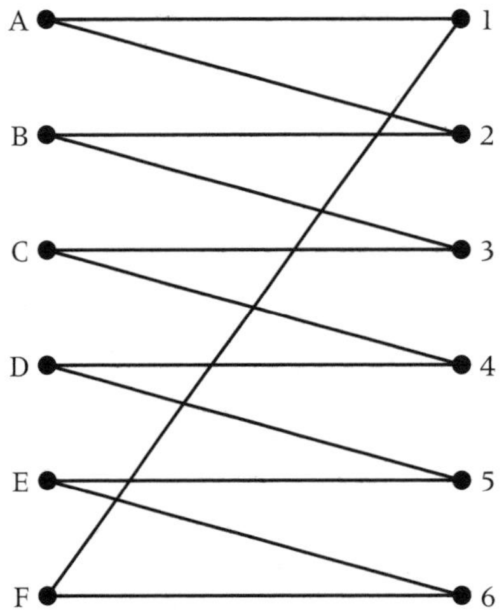

Initially A is matched to 2, B to 3, C to 4 and F to 6.
Explain why an alternating path will have to be applied twice to obtain a complete match. Apply the algorithm twice and list your complete match.

MIXED EXERCISE

1 A group of five pupils have to colour in some pictures. There are five coloured crayons available, orange (O), red (R), yellow (Y), green (G) and purple (P). The five pupils have each told their teacher their first and second choice of coloured crayons.

Pupil	First choice	Second choice
Alison (A)	orange	yellow
Brian (B)	orange	red
Carly (C)	yellow	purple
Danny (D)	red	purple
Emma (E)	purple	green

(a) Show this information on a bipartite graph.

(b) Initially the teacher gives pupils A, C, D and E their first choice of crayons. Demonstrate, by using an algorithm from this initial matching, how the teacher can give each pupil either their first or second choice of coloured crayons. [A]

2 Six children are going to eat some fruit pastilles. There are six pastilles: blackcurrant (B), orange (O), plum (P), raspberry (R), strawberry (S) and tango (T). The six children will eat only certain flavours.

Name	Flavours
Alison (A)	B
Chris (C)	O, P, S
Derek (D)	B, P, R, S
Eddie (E)	O
Freda (F)	P, T
Gemma (G)	O, R, S

(a) Show this information on a bipartite graph.

(b) Initially, Chris chooses orange, Derek chooses blackcurrant, Freda chooses plum and Gemma chooses a strawberry pastille.

Demonstrate, by using an alternating path from this initial matching, how each child can get a pastille that they will eat. [A]

6

3 Four people, A, B, C and D are to be matched to four tasks, 1, 2, 3 and 4.
A bipartite graph showing the possible allocation of people to jobs is shown below (left).
An initial matching is shown below (right).

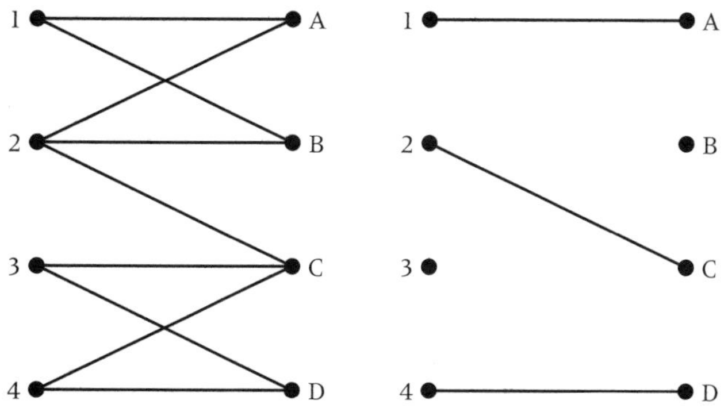

(a) Write down an adjacency matrix that represents the bipartite graph shown above (left).

(b) There are four distinct alternating paths that can be generated from the initial matching shown above (right).

One of the paths is 3 – C ≠ 2 – A ≠ 1 – B,

which produces the complete matching 1 – B, 2 – A, 3 – C, 4 – D.

(i) Use the maximum matching algorithm from the initial matching to find another maximum matching, listing the complete matching generated.

(ii) Find the remaining two alternating paths and list the complete matchings generated in each case. [A]

4 The adjacency matrix below shows the availability of five workers, A, B, C, D and E, to complete five tasks, 1, 2, 3, 4 and 5.

	1	2	3	4	5
A	1	0	1	0	0
B	1	1	0	0	0
C	0	0	1	0	1
D	0	1	0	0	1
E	0	0	0	1	1

(a) Represent this information on a bipartite graph.

(b) Initially, worker A is assigned to task 1, worker C to task 3, worker D to task 2 and worker E to task 5.

Demonstrate, by using an algorithm from this initial matching, how each worker can be assigned to a task for which they are available. [A]

5 (a) Draw a bipartite graph representing the following adjacency matrix.

	1	2	3	4	5
A	1	0	1	0	0
B	0	0	1	1	0
C	0	1	0	1	0
D	0	0	1	1	0
E	1	1	0	0	1

(b) Given that initially A is matched to 3, B is matched to 4 and E is matched to 1, use the maximum matching algorithm, from this initial matching, to find a complete matching. List your complete matching. [A]

6 A group of five students are applying to five different universities. The students wish to visit the universities on 14th October but their teacher insists that no more than one student be allowed to visit the same university on that day. They list the two universities that they would like to visit.

Student	First choice	Second choice
Andrew (A)	Cambridge	Leeds
Joanne (J)	Cambridge	Durham
Rick (R)	Leeds	Bristol
Sarah (S)	Durham	Bristol
Tom (T)	Bristol	Oxford

(a) Draw a bipartite graph linking the students to their chosen two universities.

(b) Initially the teacher gives Andrew, Rick, Sarah and Tom their first choices of university.

Demonstrate, by using an algorithm from this initial matching, how the teacher can allocate each pupil to attend either their first or second choice of university. [A]

7 Six people, A, B, C, D, E and F, are to be matched to six tasks, 1, 2, 3, 4, 5 and 6. The following table shows the tasks that each of the people is able to undertake.

Person	Task(s)
A	1, 2, 3
B	1, 2, 5
C	2, 3, 4
D	3, 4, 6
E	3
F	2

(a) Show this information on a bipartite graph.

(b) The following is the initial matching: A – 2, B – 1, C – 3, D – 4.

Demonstrate, by using an algorithm from this initial matching, how each person can be allocated a task that they can undertake. [A]

Key point summary

1 A **matching** is a set of edges that have no vertices in common. p112

2 A **maximum matching** is a matching that contains the largest possible number of edges. p112

3 A **complete matching** is where each vertex in one set is matched to a different vertex in another set. p112

4 An **alternating path algorithm** is a method for improving an initial matching to find a maximum matching. p114

Test yourself

What to review

1 Represent the information below by: *Sections 6.1, 6.2, 6.3*

(a) a bipartite graph, and

(b) an adjacency matrix.

A 1, 3
B 2, 3, 4
C 1, 4
D 2, 4

(c) If initially A is matched to 1, B to 4 and D to 2, use an alternating path to find a complete match.

(d) Write down a complete match.

2 Give an example of a bipartite graph that has two sets of vertices but does not have a complete match. *Sections 6.1, 6.3*

3 A bipartite graph is shown below. *Sections 6.1, 6.2, 6.3*

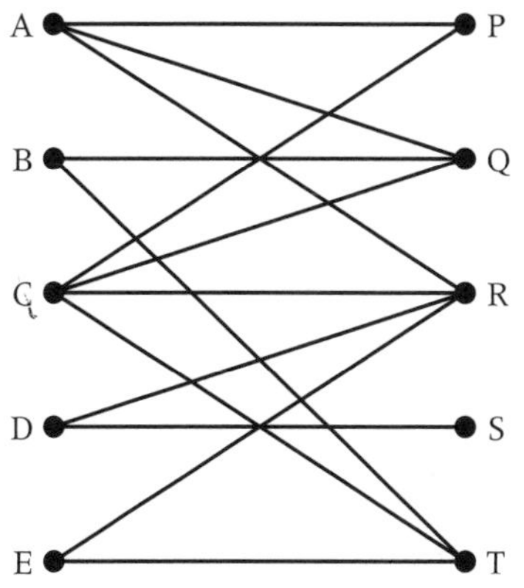

Initially A is matched to P, C is matched to Q and D is matched to S.
Show an alternating path that can be generated from this initial match.

Test yourself (continued) — What to review

4 Represent the information below as a bipartite graph. *Section 6.1*

	P	Q	R	S	T
A	0	0	1	1	0
B	1	0	0	0	1
C	1	1	0	0	0
D	1	0	0	0	1
E	0	1	0	0	1

5 Three people, A, B and C, are to be matched to three tasks, 1, 2 and 3. Each person can be matched to any of the three tasks. *Sections 6.1, 6.3*

(a) Find the maximum number of distinct matchings.

(b) Find the maximum number of distinct matchings if n people are matched to n tasks.

Test yourself ANSWERS

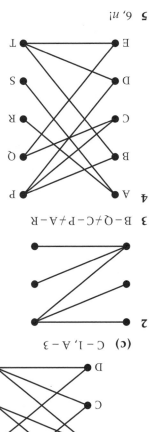

1 (a)

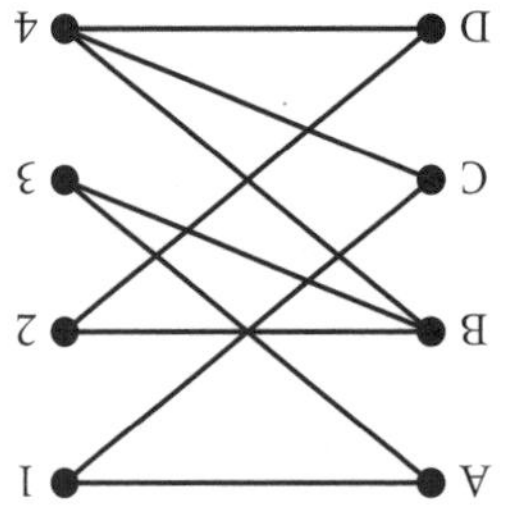

(b)

	1	2	3	4
A	1	0	1	0
B	0	1	1	1
C	1	0	0	1
D	0	1	0	1

(c) C − 1, A − 3

(d) A3, B4, C1, D2

2

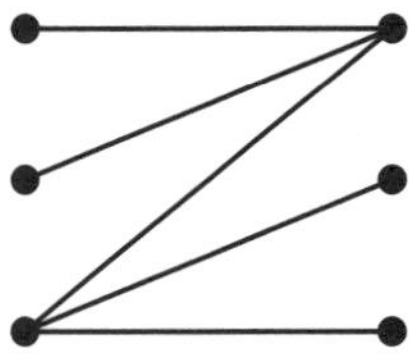

3 B−Q≠C−P≠A−R

4

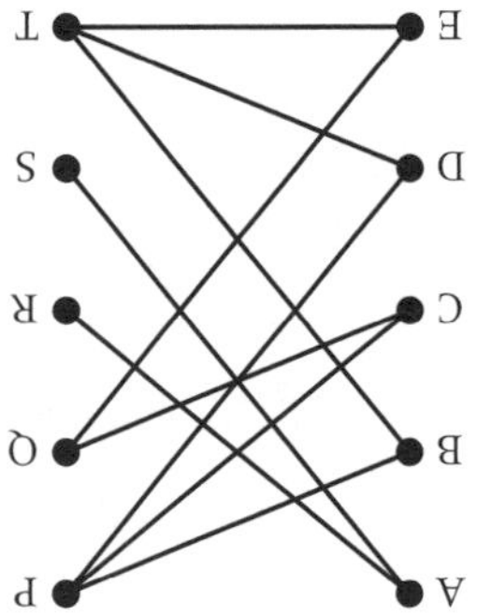

5 6, $n!$

CHAPTER 7

Sorting algorithms

Learning objectives

After studying this chapter you should be able to apply a:

- bubble-sort algorithm
- shuttle-sort algorithm
- Shell-sort algorithm
- quick-sort algorithm.

7.1 Introduction

Sorting algorithms are of greater importance as the development of computer databases increases. As data is collected, e.g. the 10-yearly census, it is essential that information can be rapidly extracted. Algorithms have been developed to speed up the analysis and retrieval of such data. This section deals with one aspect of this: sorting a list into alphabetical order or sorting numbers into numerical order.

We consider four different sorting algorithms here, and although no single algorithm is the 'best', each of the four algorithms has its own merits.

7.2 Bubble sort

One method of sorting numbers into numerical order is the **bubble sort**. The algorithm is based on comparing successive pairs of numbers. Initially the largest number is located at the end of the list, then the next largest, etc. until the whole list is in order. The following example shows this process.

Rearrange the list 18, 5, 12, 9, 3, 7, 8, 2 into ascending order.

Original list	(18, 5,) 12, 9, 3, 7, 8, 2	18 > 5, so swap the numbers
After first comparison	5, (18, 12,) 9, 3, 7, 8, 2	18 > 12, so swap the numbers
After second comparison	5, 12, (18, 9,) 3, 7, 8, 2	18 > 9, so swap the numbers
After third comparison	5, 12, 9, (18, 3,) 7, 8, 2	18 > 3, so swap the numbers

After fourth comparison	5, 12, 9, 3, (18, 7,) 8, 2
After fifth comparison	5, 12, 9, 3, 7, (18, 8,) 2
After sixth comparison	5, 12, 9, 3, 7, 8, (18, 2)
After seventh comparison	5, 12, 9, 3, 7, 8, 2, 18
Sorted list after first pass	5, 12, 9, 3, 7, 8, 2, 18

18 > 7, so swap the numbers

18 > 8, so swap the numbers

18 > 2, so swap the numbers

After the first pass the largest number, in this case 18, is located in the correct place at the end of the list. The second pass ensures that the second largest number is in the correct place. To be sure of getting all of the numbers in the correct place, there needs to be up to a maximum of $(n - 1)$ passes.

We now apply the algorithm to our list after the first pass, i.e. to the first seven numbers, because the eighth number is now in the correct place. We continue in this way until all numbers are sorted into ascending order. We do not need to list all comparisons at each stage, just state the list after each complete pass.

The set of numbers after each pass is as follows:

You will notice from the original set of comparisons why this is called a bubble-sort algorithm: there are bubbles drawn round each pair of numbers as you compare them.

After the first pass	5,	12,	9,	3,	7,	8,	2,	18
Second pass	5,	9,	3,	7,	8,	2,	12,	18
Third pass	5,	3,	7,	8,	2,	9,	12,	18
Fourth pass	3,	5,	7,	2,	8,	9,	12,	18
Fifth pass	3,	5,	2,	7,	8,	9,	12,	18
Sixth pass	3,	2,	5,	7,	8,	9,	12,	18
Seventh pass	2,	3,	5,	7,	8,	9,	12,	18

If there are eight numbers in the list, it may require seven complete passes to put all the numbers in the correct order.

The number of swaps and comparisons at each pass is shown in the table below.

Pass	**1st**	**2nd**	**3rd**	**4th**	**5th**	**6th**	**7th**	**Total**
Comparisons	7	6	5	4	3	2	1	28
Swaps	7	5	4	2	1	1	1	21

If we were writing this list in descending order, the difference would be that after the first pass the 2 would be in the correct position, after the second pass the 3 would be in the correct position, etc.

Exactly the same principles can be used to sort a set of letters into alphabetical order and this is illustrated in Worked example 7.1.

Worked example 7.1

Use a bubble sort to rearrange the following names into alphabetical order: John (J), Bill (B), Ian (I), Rick (R), Ahmed (A), Wasim (W) and Eddie (E).

Solution

Original list	J, B, I, R, A, W, E
After first pass	B, I, J, A, R, E, W
After second pass	B, I, A, J, E, R, W
After third pass	B, A, I, E, J, R, W
After fourth pass	A, B, E, I, J, R, W
After fifth pass	A, B, E, I, J, R, W

Notice: The last two rows are identical, checking that no more swaps are needed.

The bubble-sort algorithm is straightforward to follow, but it is not often used as it is not the most efficient. The worst case scenario would be sorting a list originally written in descending order into ascending order. In this case the actual number of comparisons needed to rewrite the list into ascending order would be $n(n - 1)/2$.
This formula is easily proven by considering that if there are n numbers and the numbers are in completely the wrong order, then it will take $(n - 1)$ comparisons on the first pass, $(n - 2)$ on the second pass, $(n - 3)$ on the third pass, etc. and the sum of this series is $n(n - 1)/2$.

7

EXERCISE 7A

1 Use a bubble sort to rearrange the numbers below into:

(a) ascending order,

(b) descending order,

showing the order after each pass in both cases.

18, 3, 45, 17, 1, 26, 43, 22, 16

2 Use a bubble sort to rearrange the following numbers into ascending order, writing down the number of comparisons and swaps after each pass:

32, 11, 3, 27, 16, 9, 23, 19

3 Use a bubble sort to rearrange the following list of flowers into alphabetical order, giving the number of comparisons and swaps after each pass:

rose, lily, carnation, aster, bluebell, dahlia, wallflower, hydrangea

7.3 Shuttle sort

A shuttle sort can be more powerful than a bubble sort as a number can move more than one place in one complete pass.

> A **shuttle sort** initially compares the first and second numbers, and orders them correctly. It then introduces the third number into the sort and locates that in the correct position, then the fourth into the correct position and so on until the list is fully correct.

If we apply the shuttle-sort algorithm to the set of numbers used in the previous example we obtain the following:

Original list 18, 5, 12, 9, 3, 7, 8, 2 compares first two numbers
After first pass 5, 18, 12, 9, 3, 7, 8, 2 compares first three numbers
After second pass 5, 12, 18, 9, 3, 7, 8, 2 etc.
After third pass 5, 9, 12, 18, 3, 7, 8, 2
After fourth pass 3, 5, 9, 12, 18, 7, 8, 2
After fifth pass 3, 5, 7, 9, 12, 18, 8, 2
After sixth pass 3, 5, 7, 8, 9, 12, 18, 2
After seventh pass 2, 3, 5, 7, 8, 9, 12, 18

It is easy to illustrate a shuttle sort by simply underlining the sets of numbers that have been compared at each pass and adding one extra number to the list for each subsequent pass. The number of swaps and comparisons at each pass is shown in the table below.

Pass	**1st**	**2nd**	**3rd**	**4th**	**5th**	**6th**	**7th**	**Total**
Comparisons	1	2	3	4	4	4	7	25
Swaps	1	1	2	3	3	3	7	20

As before with the bubble sort, if the list is written in descending order then the shuttle sort does not reduce the number of comparisons because as each number is added to the list it needs to be compared with all the previous numbers.

Worked example 7.2

Use a shuttle sort to rearrange the following names into alphabetical order: John (J), Bill (B), Ian (I), Rick (R), Ahmed (A), Wasim (W) and Eddie (E).

Solution

Original list J, B, I, R, A, W, E
After first pass B, J, I, R, A, W, E
After second pass B, I, J, R, A, W, E
After third pass B, I, J, R, A, W, E
After fourth pass A, B, I, J, R, W, E

After fifth pass A, B, I, J, R, W, E
After sixth pass A, B, E, I, J, R, W

EXERCISE 7B

1 Use a shuttle sort to rearrange the numbers below into:
   (a) ascending order,
   (b) descending order,
   showing the order after each pass in both cases.
   18, 3, 45, 17, 1, 26, 43, 22, 16

2 Use a shuttle sort to rearrange the following numbers into ascending order, writing down the number of comparisons and swaps after each pass:
   32, 11, 3, 27, 16, 9, 23, 19

3 Use a shuttle sort to rearrange the following list of flowers into alphabetical order, giving the number of comparisons and swaps after each pass:
   rose, lily, carnation, aster, bluebell, dahlia, wallflower, hydrangea

7.4 Shell sort

If a large set of numbers is to be sorted, both the bubble sort and the shuttle sort are very time-consuming.
In 1959 a mathematician called D.L. Shell devised the Shell sort as a more efficient way of sorting large sets of numbers. The algorithm splits the set of numbers into smaller sets, sorting them and then merging the sorted subsets together.

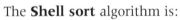

The **Shell sort** algorithm is:

Step 1 Divide a list into $n/2$ sublists, ignoring any remainders (a sublist is part of a list).

Step 2 Shuttle sort each sublist.

Step 3 Merge the sorted sublists.

Step 4 Divide the number of sublists by 2 and repeat steps 2 and 3 until there is only one sublist.

Although this algorithm may seem awkward to follow, the application is demonstrated easily on the same worked example:

18, 5, 12, 9, 3, 7, 8, 2

As there are eight numbers, initially there will be $8/2 = 4$ sublists. The sublists will comprise:

the first number and the fifth number
the second number and the sixth number
the third number and the seventh number
the fourth number and the eighth number

This means that we will compare the 18 and the 3, the 5 and the 7, the 12 and the 8, the 9 and the 2 (see step 1 below).

The shuttle sort of this is obvious and having rearranged the numbers into order and then merged the four sublists together, after the first complete pass our new list will be:

3, 5, 8, 2, 18, 7, 12, 9

(see steps 2 and 3 below).

On the first pass there were four sublists so on the second pass there will be 4/2 = 2 sublists, each containing four numbers. The two sublists will contain:

the first, the third, the fifth and the seventh numbers
the second, the fourth, the sixth and the eighth numbers

These two sublists are sorted using the shuttle sort and then merged together. This gives a new list after the second pass of:

3, 2, 8, 5, 12, 7, 18, 9

There were two sublists at the second pass, so now dividing 2 by 1 gives one sublist. This means that we now shuttle sort this list. As you can see from the values we currently have, it is far easier to shuttle sort this list than it was the original list. Applying the shuttle sort to this sublist gives the numbers in the correct order. Writing the list at each step gives:

Initial list	18,	5,	12,	9,	3,	7,	8,	2	
Four sublists	18				3				
		5				7			
			12				8		
				9				2	1st pass
Sort	3				18				
		5				7			
			8				12		
				2				9	
Merge	3,	5,	8,	2,	18,	7,	12,	9	
Two sublists	3		8		18		12		
		5		2		7		9	
Sort	3		8		12		18		2nd pass
		2		5		7		9	
Merge	3,	2,	8,	5,	12,	7,	18,	9	
One sublist	3,	2,	8,	5,	12,	7,	18,	9	3rd pass
Sort	2,	3,	5,	7,	8,	9,	12,	18	

The number of swaps and comparisons at each pass is shown in the table below.

Pass	1st	2nd	3rd	4th	5th	6th	7th	Total
Comparisons	4	6	12					22
Swaps	3	2	7					12

Worked example 7.3

Use a Shell sort to rearrange the following names into alphabetical order: John (J), Bill (B), Ian (I), Rick (R), Ahmed (A), Wasim (W) and Eddie (E).

Solution

Original list	J,	B,	I,	R,	A,	W,	E
7/2 = 3 sublists	J			R			E
		B			A		
			I			W	
Sort and merge	E	A	I	J	B	W	R
3/2 = 1 sublist	E,	A,	I,	J,	B,	W,	R
Sort and merge	A,	B,	E,	I,	J,	R,	W

EXERCISE 7C

1 Use a Shell sort to rearrange the numbers below into:
- **(a)** ascending order,
- **(b)** descending order,

showing the order after each pass in both cases.

18, 3, 45, 17, 1, 26, 43, 22, 16

2 Use a Shell sort to rearrange the following numbers into ascending order, writing down the number of comparisons and swaps after each pass:

32, 11, 3, 27, 16, 9, 23, 19

3 Use a Shell sort to rearrange the following list of flowers into alphabetical order, giving the number of comparisons and swaps after each pass:

rose, lily, carnation, aster, bluebell, dahlia, wallflower, hydrangea

7.5 Quick sort

This algorithm was introduced in 1962 by Hoare as a quick means of sorting a set of numbers. In many texts the quick-sort algorithm involves locating the middle number in the list and using this number as a pivot and creating new sublists. There is no reason, however, why using the middle number in the list as the pivot is better and it is a difficult concept to follow. It is far easier to take the first number in each list or sublist as the pivot.

The **quick sort** algorithm is:

Step 1 Use the first number in each list as the pivot.

Step 2 Create new sublist(s) by writing the numbers lower than the pivot to the left and higher numbers to the right (do not change the order of numbers in the sublist).

Step 3 Go to Step 1 and repeat the process on each sublist until each sublist only has one element.

Sometimes the sublists will be empty.

We will apply this algorithm to the same set of numbers as used earlier.
(Notation: underlined pivots are currently being used and boxed pivots have been used)

Original list	$\underline{18}$, 5, 12, 9, 3, 7, 8, 2	pivot is 18

(Note: as all numbers are less than 18, there is only one sublist)

After first pass	$\underline{5}$, 12, 9, 3, 7, 8, 2, $\boxed{18}$	pivot is 5
Second pass	$\underline{3}$, 2, $\boxed{5}$, $\underline{12}$, 9, 7, 8, $\boxed{18}$	pivots are 3 and 12
Third pass	$\underline{2}$, $\boxed{3}$, $\boxed{5}$, $\underline{9}$, 7, 8, $\boxed{12}$, $\boxed{18}$	pivots are 2 and 9
Fourth pass	$\boxed{2}$, $\boxed{3}$, $\boxed{5}$, $\underline{7}$, 8, $\boxed{9}$, $\boxed{12}$, $\boxed{18}$	pivot is 7
Fifth pass	$\boxed{2}$, $\boxed{3}$, $\boxed{5}$, $\boxed{7}$, $\underline{8}$, $\boxed{9}$, $\boxed{12}$, $\boxed{18}$	pivot is 8
Sixth pass	$\boxed{2}$, $\boxed{3}$, $\boxed{5}$, $\boxed{7}$, $\boxed{8}$, $\boxed{9}$, $\boxed{12}$, $\boxed{18}$	

At this point each sublist only contains one element and hence the list has been sorted. The number of swaps and comparisons at each pass is shown in the table below.

Pass	1st	2nd	3rd	4th	5th	6th	7th	Total
Comparisons	7	6	4	2	1			20
Swaps	7	2	4	2	0			15

Worked example 7.4

Use a quick sort to rearrange the following names into alphabetical order: John (J), Bill (B), Ian (I), Rick (R), Ahmed (A), Wasim (W) and Eddie (E).

Solution

Original list	$\underline{J}$, B, I, R, A, W, E,	pivot is J
After first pass	$\underline{B}$, I, A, E, $\boxed{J}$, $\underline{R}$, W	pivots are B and R
After second pass	$\underline{A}$, $\boxed{B}$, $\underline{I}$, E, $\boxed{J}$, $\boxed{R}$, $\underline{W}$	pivots are A, I and W
After third pass	$\boxed{A}$, $\boxed{B}$, $\underline{E}$, $\boxed{I}$, $\boxed{J}$, $\boxed{R}$, $\boxed{W}$	pivot is E
After fourth pass	$\boxed{A}$, $\boxed{B}$, $\boxed{E}$, $\boxed{I}$, $\boxed{J}$, $\boxed{R}$, $\boxed{W}$	

EXERCISE 7D

1 Use a quick sort to rearrange the numbers below into:

(a) ascending order,

(b) descending order,

showing the order after each pass in each case.

18, 3, 45, 17, 1, 26, 43, 22, 16

2 Use a quick sort to rearrange the following numbers into ascending order, writing down the number of comparisons and swaps after each pass:

32, 11, 3, 27, 16, 9, 23, 19

3 Use a quick sort to rearrange the following list of flowers into alphabetical order, giving the number of comparisons and swaps after each pass:

rose, lily, carnation, aster, bluebell, dahlia, wallflower, hydrangea

7.6 Efficiency of methods

The table below indicates the number of comparisons and the number of swaps needed using each of the four methods.

Pass	Bubble		Shuttle		Shell		Quick	
	Comparisons	Swaps	Comparisons	Swaps	Comparisons	Swaps	Comparisons	Swaps
1st	7	7	1	1	4	3	7	7
2nd	6	5	2	1	6	2	6	2
3rd	5	4	3	2	12	7	4	4
4th	4	2	4	3			2	2
5th	3	1	4	3			1	0
6th	2	1	4	3				
7th	1	1	7	7				
Total	28	21	25	20	22	12	20	15

On the set of numbers we have used, you can see that the quick sort requires the fewest comparisons and the Shell sort requires the fewest swaps. The bubble sort is the least efficient. The efficiency of the methods does depend on the order of the original list of numbers.

7

MIXED EXERCISE

1 (a) Use a bubble-sort algorithm to rearrange the following numbers into ascending order, showing the new arrangement after each pass:

4, 7, 13, 26, 8, 15, 6, 56

(b) Find the maximum number of comparisons needed to rearrange a list of eight numbers into ascending order. [A]

2 Use the quick-sort algorithm to rearrange the following list of flowers into alphabetical order:

rose (R), iris (I), wallflower (W), dahlia (D), pansy (P), lobelia (L), azalea (A)

Indicate the entries that you have used as pivots. [A]

3 Use a Shell-sort algorithm to rearrange the following numbers into ascending order, showing the new arrangement after each pass:

14, 27, 23, 36, 18, 25, 16, 66

4 There are five mathematicians who are members of a committee: Newton (N), Euler (E), Descartes (D), Pythagoras (P) and Archimedes (A).

Use a bubble-sort algorithm to rearrange these names into alphabetical order, showing the new arrangement after each comparison. [A]

5 (a) Use the quick-sort algorithm to rearrange the following numbers into order, showing the new arrangement at each stage. Take the first number in any list as the pivot.

9, 5, 7, 11, 2, 8, 6, 17 [A]

(b) A shuttle-sort algorithm is to be used to rearrange a list of numbers into order.

(i) Find the maximum number of comparisons that would be needed to be certain that a list of eight numbers was in order.

(ii) Find, in a simplified form, an expression for the maximum number of comparisons that would be needed to be certain that a list containing n numbers was in order. [A]

6 Use the quick-sort algorithm to rearrange the following list of numbers into ascending order:

63, 32, 70, 26, 59, 41, 17

Indicate entries that you have used as pivots. [A]

7 (a) Use the quick-sort algorithm to rearrange the following colours into alphabetical order, showing the new arrangement at each stage.

pink, black, red, green, orange, white

(b) Find the maximum number of comparisons that could be needed to sort a list of six words into alphabetical order.

(c) Find an expression for the maximum number of comparisons that could be needed to sort a list of n words into alphabetical order. [A]

Key point summary

1	A **bubble sort** rearranges a list by finding the largest element on each pass.	*p123*
2	A **shuttle sort** adds a new number to a list, placing it in its correct position on each pass.	*p126*
3	A **Shell sort** divides a list into smaller sublists, called sorts, to produce a new list after each pass.	*p127*
4	A **quick sort** divides a list using pivots until each sublist contains only one element.	*p129*

Test yourself

Test yourself	What to review
1 Rearrange the list below into ascending order using: **(a)** a bubble sort, **(b)** a quick sort. 12, 5, 17, 9, 11, 2 For each method write down the number of comparisons and swaps.	*Sections 7.2, 7.5*
2 Rearrange the list below into alphabetical order using: **(a)** a shuttle sort, **(b)** a shell sort. R, M, A, S, B, L, Z, P	*Sections 7.3, 7.4*
3 Use a bubble sort to rearrange the following list into ascending order: 12, −3, 24, 11, −2, 17, −5	*Section 7.2*
4 Use a Shell sort to rearrange the following set of numbers into ascending order: 23, 11, −6, 19, 32, 5, −8, 13	*Section 7.4*

Test yourself ANSWERS

1 (a)

12	5	17	9	11	2	C	S
5	12	9	11	2	17	5	4
5	9	11	2	12	17	4	3
5	9	2	11	12	17	3	1
5	2	9	11	12	17	2	1
2	5	9	11	12	17	1	1

(b)

12	5	17	9	11	2	C	S
5	9	11	2	12	17	5	4
2	5	9	11	12	17	3	1
2	5	9	11	12	17	1	0

2 (a)

R	M	A	S	B	L	Z	P
M	R	A	S	B	L	Z	P
A	L	B	M	R	P	Z	S
A	M	R	S	B	L	Z	P
A	M	R	S	B	L	Z	P
A	B	M	R	S	L	Z	P
A	B	L	M	R	S	Z	P
A	B	L	M	R	S	Z	P
A	B	L	M	P	R	S	Z

(b)

R	M	A	S	B	L	Z	P
B	L	A	P	R	M	Z	S
A	B	L	M	P	R	S	Z

3

12	−3	24	11	−2	17	−5
−3	12	11	−2	17	−5	24
−3	11	−2	12	−5	17	24
−3	−2	11	−5	12	17	24
−3	−2	−5	11	12	17	24
−3	−5	−2	11	12	17	24
−5	−3	−2	11	12	17	24

4

23	11	−6	19	32	5	−8	13
23	5	−8	13	32	11	−6	19
−8	5	−6	11	23	13	32	19
−8	−6	5	11	13	19	23	32

CHAPTER 8

Algorithms

Learning objectives

After studying this section you should understand:

- what an algorithm is
- flow diagrams
- how to trace flow diagrams
- how to trace a series of instructions written in pseudo English
- the meaning of correctness and stopping conditions.

8.1 Introduction

The word algorithm comes from the Persian mathematician Mohammed Al-Khowarizmi. His most famous work was Al-jabr wal-muquabalah, which is the science of the equations and ultimately gave us the words algebra and algorithm.

An algorithm is a sequence of instructions that allows anyone to solve a problem. We have already met algorithms earlier in the book, e.g. Prim's, Kruskal's and Dijkstra's. Although we normally associate the word algorithm with solving mathematical problems, it is quite valid in everyday life, e.g. how do you make a cup of tea, change the wheel on a car?

It is essential that any algorithm has a set of precise instructions, so that there is no ambiguity whatsoever as to what the instructions are and the order in which they are to be carried out. If you are changing a wheel, it is essential that the car is jacked up before the old wheel is removed, which must be before the new wheel is put in place, etc.

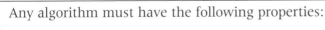

Any algorithm must have the following properties:

- there must be a finite number of instructions
- each stage must be defined precisely
- each instruction must be precise
- each answer must depend only on the values input to solve a particular problem
- the algorithm must work for any set of values that are input.

Worked example 8.1

Write down a set of instructions for making vegetable stir fry.

Solution

Step 1
Wash the broccoli.
Peel and chop the garlic.
Peel and grate the ginger.
Peel and slice the carrots.
Deseed and chop the peppers.

Step 2
Heat the oil in the wok until hot.
Add the garlic and ginger and stir-fry for 1 minute.
Add the broccoli, peppers and carrots and stir fry for 5 minutes.

Step 3
Add the stock to the wok and cook for a further 5 minutes.
Stir in the oyster sauce.

EXERCISE 8A

1 Write down a set of instructions for making a cup of tea.

2 Write down a set of instructions for changing a wheel on a car.

3 Write down a set of instructions for finding your way from home to school each day.

In all of these algorithms there will be some decisions that will need to be made and the answers to those decisions will determine the next stage of the algorithm.

For example, when changing the wheel on a car we must remove the wheel nuts so we will need an instruction to remove the wheel nuts, then a question: have we removed all the wheel nuts? If the answer is no, then we will go back and remove another wheel nut. If the answer is yes, then we will remove the wheel. It is always interesting at this stage to check that if you followed a set of instructions whether or not the outcome will be exactly as desired.

A lot of furniture these days arrives in flat pack form, with a series of instructions. Companies are often accused of not making the instructions clear enough so that anybody can make the furniture and it will all look the same. It is therefore essential that when any algorithm has been written it is tested to ensure that it gives the desired result with no ambiguity, in a finite time and regardless of who will actually apply the algorithm.

There are two standard methods of representing algorithms:

- flow diagrams,
- a series of instructions written in pseudo English.

8.2 Flow diagrams

In pure mathematics the formula for solving quadratic equations of the form $ax^2 + bx + c = 0$ is:

$$x = \frac{-b \pm \sqrt{(b^2 - 4ac)}}{2a}$$

This is an algorithm.

The algorithm may be a flow chart or diagram, as illustrated below.

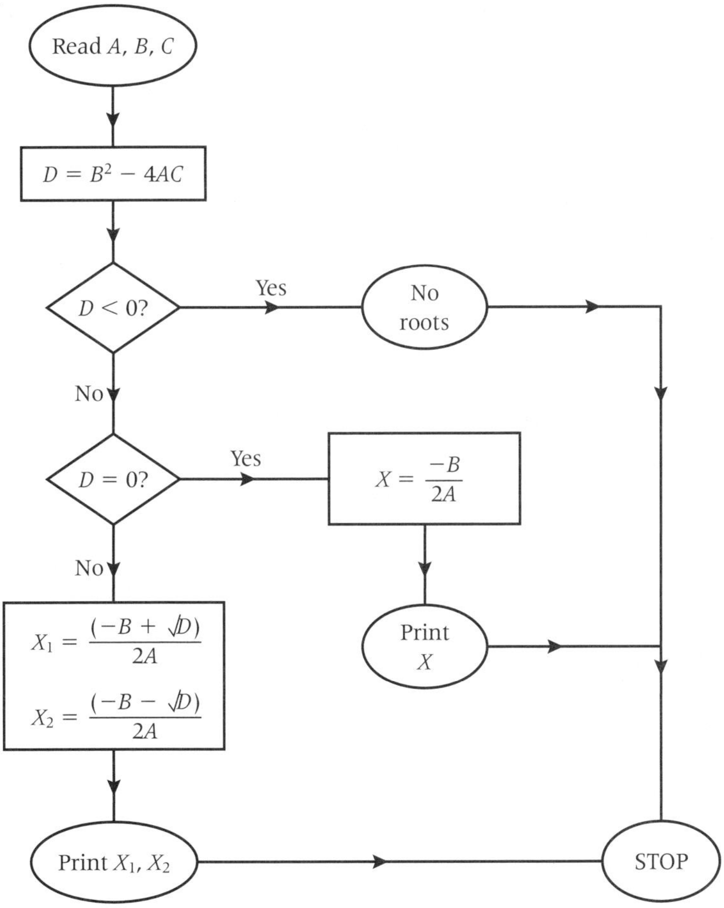

When using flow charts:

- oval boxes are used for starting, stopping, inputting or outputting data,
- square boxes are used for calculations or instructions,
- diamond-shaped boxes are used for decisions.

Worked example 8.2

Using the flow diagram below, find the output in the case where $N = 18$.

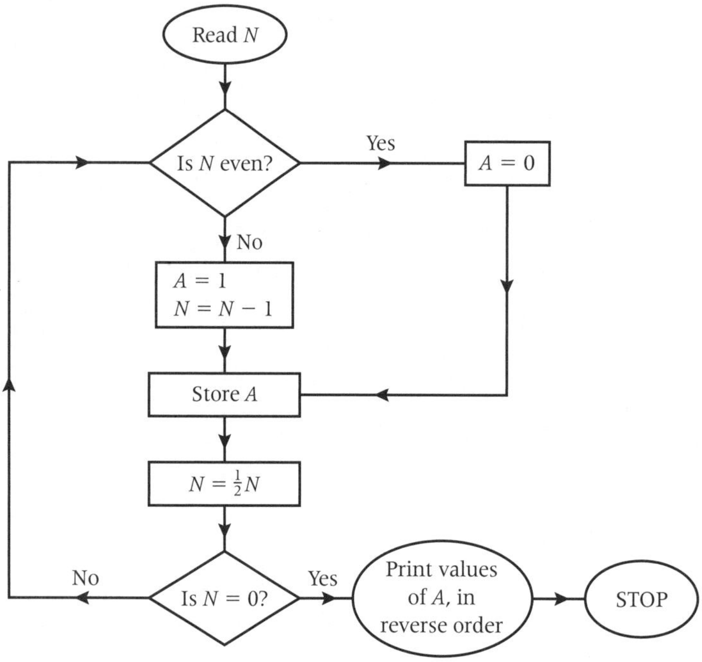

Solution

After each pass around the flow diagram, the values of N, A and the printed numbers are as shown in the table below.

Pass	**0**	**1**	**2**	**3**	**4**	**5**
N	18	9	8	4	2	1
A		0	1	0	0	1

The printed result is 10 010.

EXERCISE 8B

1 Trace the algorithm below for the case where $A = 1$.

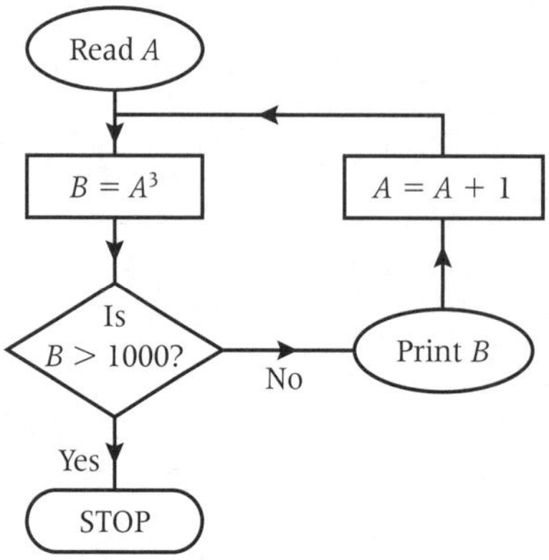

2 A flow diagram is shown below.

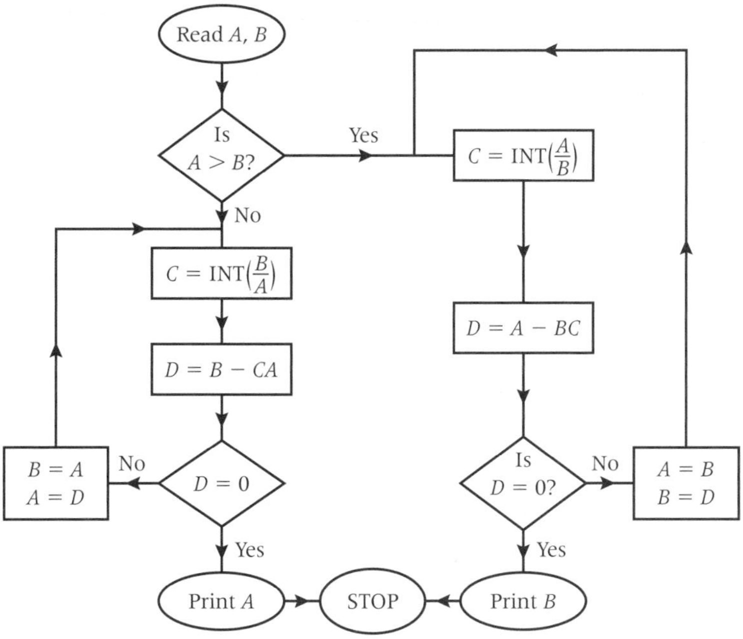

What is the output of the flow diagram where:

(a) $A = 20, B = 24$

(b) $A = 72, B = 60$

3 A flow diagram for multiplying two numbers is shown below. Trace the flow diagram in the case where $A = 27$ and $B = 15$. (*Note*: this algorithm for multiplying two numbers together is called the Russian peasant's algorithm.)

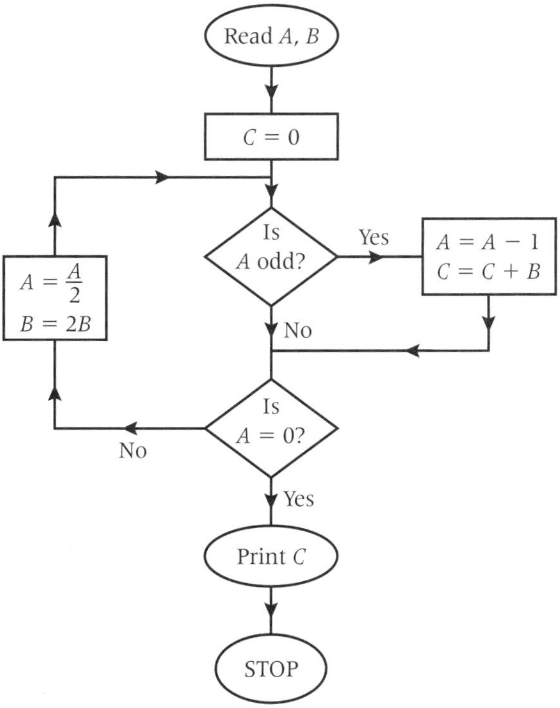

8.3 Set of instructions written in pseudo English

The formula for solving quadratic equations of the form $ax^2 + bx + c = 0$ is:

$$x = \frac{-b \pm \sqrt{(b^2 - 4ac)}}{2a}$$

This algorithm can be represented by the following set of instructions.

```
LINE 10   Input A, B, C
LINE 20   Let D = B*B − 4*A*C
LINE 30   Let X1 = (−B + √D)/(2*A)
LINE 40   Let X2 = (−B − √D)/(2*A)
LINE 50   Print X1, X2
LINE 60   Stop
```

> When an algorithm is written as a series of instructions:
>
> - the instructions are written in a simplistic form and are prefixed with line numbers,
> - the line numbers themselves do not affect the working of an algorithm, they are an indication as to the order of the instructions.

Line numbers are normally written in multiples of 10 so that if there is any mistake or any new lines need to be added, the existing lines will not need to be renumbered. An insert can be added at a later stage with an appropriate line number.

The current algorithm does not have a test to see if the equation has real roots. To amend this algorithm we would add a new LINE 25:

```
LINE 25   If D < 0 then goto LINE 60
```

Further lines could be added if a printout of the fact that there were no real roots was to be included:

```
LINE 25   If D < 0 then goto LINE 54
LINE 52   Goto LINE 60
LINE 54   Print 'No real roots'
```

Worked example 8.3

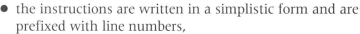

Trace the following algorithm to find the output in the case where $N = 18$.

```
LINE 10   Input N
LINE 20   Let K = 0
LINE 30   If N/2 = INT(N/2) then goto LINE 70
LINE 40   Let A = 1
LINE 50   Let N = N − 1
```

Here is an alternative way of writing this algorithm. The only difference is that lines that refer to labels (e.g. Label *X*) have been inserted. These labels are simply milestones in the algorithm. So instead of writing 'goto LINE 30' you can write 'goto *X*'. This is

LINE 60   Goto LINE 80
LINE 70   Let $A = 0$
LINE 80   Let $B(K) = A$
LINE 90   Let $K = K + 1$
LINE 100   Let $N = N/2$
LINE 110   If $N > 0$ then goto LINE 30
LINE 120   For $L = K - 1$ to 0 step -1
LINE 130   Print $B(L)$
LINE 140   Next L
LINE 150   Stop

Solution

When tracing the instructions, the values of the variables at each line are as shown in the trace table below.

Line number	N	K	A	$B(K)$	L	$B(L)$
10	18					
20		0				
70			0			
80				0		
90		1				
100	9					
40			1			
50	8					
80				1		
90		2				
100	4					
70			0			
80				0		
90		3				
100	2					
70			0			
80				0		
90		4				
100	1					
40			1			
50	0					
80				1		
90		5				
100	0					
120 & 130					4	1
120 & 130					3	0
120 & 130					2	0
120 & 130					1	1
120 & 130					0	0

useful if you need to make changes that affect the line numbering.

LINE 10   Input N
LINE 20   Let $K = 0$
LINE 30   Label X
LINE 40   If $N/2 = \text{INT}(N/2)$ then goto Y
LINE 50   Let $A = 1$
LINE 60   Let $N = N - 1$
LINE 70   Goto Z
LINE 80   Label Y
LINE 90   Let $A = 0$
LINE 100   Label Z
LINE 110   Let $B(K) = A$
LINE 120   Let $K = K + 1$
LINE 130   Let $N = N/2$
LINE 140   If $N > 0$ then goto X
LINE 150   For $L = K - 1$ to 0 step -1
LINE 160   Print $B(L)$
LINE 170   Next L
LINE 180   Stop

On a flow chart it is easy to see by simply drawing the arrows how we may have to go around a set of instructions. However, if we are using each sentence as a statement, rather than in a flow diagram, then there are two normal methods that we use for going around a loop.

An IF THEN statement

This will ask a question and IF something is true THEN you will be directed to a line that may not be in correct numerical order.

A FOR loop

This applies a loop for a set of number of occasions, e.g. we may want to produce a number of multiplication tables and therefore we will want to trace a set of instructions FOR an exact number of times.

The relative merits of these two methods are that if we know the exact number of occasions round the loop then we will use a FOR loop, but if we are unsure as to the number of occasions around the loop, we will use the decision IF THEN.

EXERCISE 8C

1 Write an algorithm that will give the two-times table from 1 to 10.

2 Write an algorithm that will give all square numbers less than 100.

3 Write an algorithm that will find all prime numbers between 10 and 50.

4 Write an algorithm that will give multiplication tables from 1 to 10.

8.4 Stopping conditions

Consider the following algorithm:

LINE 10 Let $X = 1$
LINE 20 $Y = X*X*X$
LINE 30 Print Y
LINE 40 $X = X + 1$
LINE 50 Goto LINE 20
LINE 60 End

This algorithm is designed to print the set of cube numbers; however, the algorithm will never reach LINE 60 and consequently will never end!

Tracing the algorithm will, at each loop, increase the value of X by 1.
To correct the algorithm a stopping condition must be included. This can be done in a number of ways.
If we wanted the first 10 cube numbers, say, then:

```
LINE 10   For X = 1 to 10
LINE 20   Y = X*X*X
LINE 30   Print Y
LINE 40   Next X
LINE 50   End
```

or

```
LINE 10   Let X = 1
LINE 20   Y = X*X*X
LINE 30   Print Y
LINE 40   X = X + 1
LINE 50   If X = 11 then goto LINE 70
LINE 60   Goto LINE 20
LINE 70   End
```

However, if we wanted to stop when the cube number reached a certain figure, 500 say, then we would have to use an IF THEN question, for example:

```
LINE 10   Let X = 1
LINE 20   Y = X*X*X
LINE 30   If Y > 500 then goto LINE 70
LINE 40   Print Y
LINE 50   X = X + 1
LINE 60   Goto LINE 20
LINE 70   End
```


EXERCISE 8D

1 Write an algorithm that will give the three-times table from 5 to 10:

(a) using a FOR loop,

(b) using an IF THEN conditional test.

2 Write an algorithm that will give all square numbers between 50 and 100:

(a) using a FOR loop,

(b) using an IF THEN conditional test.

3 Write an algorithm that will give multiplication tables from 1 to 10:

(a) using a FOR loop,

(b) using an IF THEN conditional test.

MIXED EXERCISE

1 (a) Trace the algorithm below.

```
LINE 1   A = 1
LINE 2   Label X
LINE 3   B = A*A*A
LINE 4   If B > 100 then goto Y
LINE 5   Print A, B
LINE 6   A = A + 1
LINE 7   Goto X
LINE 8   Label Y
LINE 9   Stop
```

(b) Explain how your trace table would change if LINES 1 and 2 were interchanged. [A]

2 The following algorithm has been written to input a set of 30 examination marks, each expressed as an integer percentage. Find the minimum mark and output the result.

```
Line
1   SET MIN = 100
2   For I = 1 to 30
3   INPUT MARK
4   IF MARK < MIN
5   THEN MIN = MARK
6   NEXT N
7   ..........................
```

(a) (i) State the purpose of line 1 of the algorithm.

(ii) There is a mistake in line 6. Write down a corrected version of this line.

(iii) The contents of line 7 are missing. Write down the contents of line 7 to ensure that the algorithm is fully complete.

(b) Show how this algorithm could be adapted if the number of examination marks to be input was unknown. [A]

3 Consider the algorithm below, which operates on two positive integers, X and Y.

```
Read X, Y
Label A
   If X > Y then X = X − Y
   If Y > X then Y = Y − X
   If X ≠ Y then GOTO A
Print X
```

Trace the algorithm in the case where $X = 24$ and $Y = 20$.

[A]

4 A student is writing a computer program to calculate part of a multiplication table.
This is the algorithm she uses:

$X = 0$, $K = 0$
For $I = 1$ to 6
For $J = 1$ to 5
$X = I*J$
$K = K + 1$
Print K, I, 'times', J, 'equals', X
Next J
Next I
End

(a) State:

(i) the purpose of the line $X = 0$, $K = 0$,

(ii) why the lines NEXT J, NEXT I are given in that order,

(iii) the purpose of the variable K.

(b) When the value of K printed is 8, find the corresponding printed values of I, J and X. [A]

5 A schoolboy is surprised to discover an old method of multiplication which requires knowledge of just one multiplication table, the two-times table. He attempts to explain the method to a friend by writing it out as an algorithm with Q and R as positive integers.

He uses two functions called DOUBLE and SPLIT.
DOUBLE has its usual meaning of 'multiply by two', so DOUBLE(7) = 14.
SPLIT means 'divide by two but ignore any halves', so SPLIT(7) = 3, SPLIT(8) = 4.

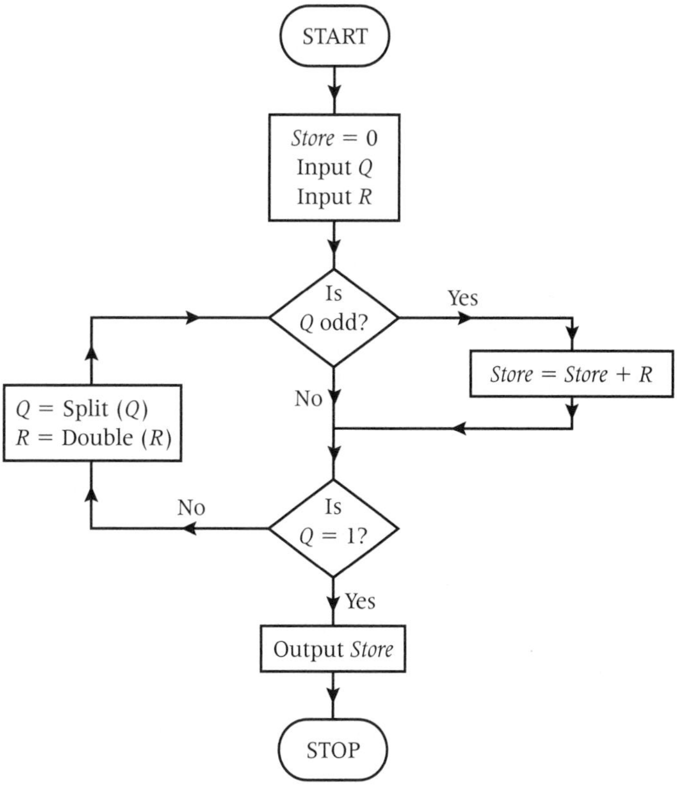

(a) Use the given algorithm to test whether the input of 7, 9 gives the expected output. Show the result of each step.

(b) Repeat to find the output of the input 4, 8.

(c) Without carrying out the algorithm, explain why the input of 10, 25 would be performed faster than the input of 25, 10. [A]

8

6 (a) Apply the following algorithm to the list of data below.

41, 23, 12, 45, 17, 11, 26, 58, 3, 24

Step 1 Choose the first element in the current list; call it P.

Step 2 For each number in the list, put numbers less than P to the left of P, and numbers greater than P to the right of P, creating sublists, each maintaining the original order of numbers.

Step 3 If every sublist has one element, then output the sublists in order. Otherwise go to step 1 and repeat for each sublist, keeping the sublists in order.

(b) Find the maximum number of comparisons needed to rearrange a list of ten numbers into ascending order using the algorithm from **(a)**.

(c) Rearrange the list of numbers in **(a)** so that the maximum number of comparisons would be required. [A]

7 A student is using the algorithm below to find the real roots of a quadratic equation.

LINE 10 Input A, B, C
LINE 20 $D = B*B - 4*A*C$
LINE 30 $X_1 = (-B + \sqrt{D})/(2*A)$
LINE 40 $X_2 = (-B - \sqrt{D})/(2*A)$
LINE 50 If $X_1 = X_2$ then goto L
LINE 60 Print 'different roots', X_1, X_2
LINE 70 Goto M
LINE 80 Label L
LINE 90 Print 'equal roots', X_1
LINE 100 Label M
LINE 110 End

(a) Trace the algorithm

(i) if $A = 1$, $B = -4$, $C = 4$,

(ii) if $A = 2$, $B = 9$, $C = 9$.

(b) **(i)** Find a set of values of A, B and C for which the algorithm would fail.

(ii) Write down additional lines to ensure that the algorithm would not fail for **any** values of A, B and C that may be input. [A]

8 A woman invests £200 on January 1 for **each** of three years in a fixed income bond that pays interest of 8 per cent per annum, the interest being added to her account at the end of each year.

The following algorithm gives the total value of her investment after the three year period.

$A = 0$
For $I = 1$ to 3
$A = A + 200$
$A = 1.08 \times A$
Next I
Print A

(a) Trace the algorithm.

(b) Explain how the algorithm could be amended to give her the value of her investment at the end of each year, after interest for that year has been added.

(c) Write a modified algorithm that would find the value, after a period of N years, of an investment of £P, invested at the start of each year, in a bond paying a constant rate of interest of R per cent per annum. [A]

9 The following algorithm is to be used on different sets of numbers.

LINE 10 Input X, Y
LINE 20 Let $A = Y$
LINE 30 Let $B = 0$
LINE 40 Let $A = A - X$
LINE 50 Let $B = B + 1$
LINE 60 If $A \leqslant X$ then goto LINE 40
LINE 70 Print A, B
LINE 80 Stop

(a) Trace the algorithm:
(i) in the case when $X = 5$ and $Y = 20$,
(ii) in the case when $X = 7$ and $Y = 29$.

(b) State the purpose of the algorithm. [A]

10 Students were investigating different numerical methods for estimating the area under the graph of $y = x^2$ between the limits $x = A$ and $x = B$.

The students used the following algorithm:

Input A, B, N
$S = 0$
$H = \dfrac{(B - A)}{N}$
$X = A$
Label 1
Procedure (calculate)
If $N > 0$ goto label 1
Area $= S$

The students used two different versions of Procedure (calculate).

Version 1	**Version 2**
$X = X + \dfrac{H}{2}$	$P = X \times X$
$Y = X \times X$	$X = X + H$
$S = S + H \times Y$	$Q = X \times X$
$X = X + \dfrac{H}{2}$	$X = X + H$
$N = N - 1$	$R = X \times X$
	$Y = P + 4 \times Q + R$
	$S = S + \dfrac{H \times Y}{3}$
	$N = N - 2$

(a) Using version 1 of Procedure (calculate), trace the algorithm for the case where $A = 1$, $B = 3$, $N = 2$.

(b) Using version 2 of Procedure (calculate), trace the algorithm for the case where $A = 1$, $B = 3$, $N = 4$. [A]

Key point summary

1 An **algorithm** is a finite set of instructions for solving a problem. *p135*

2 Algorithms can be represented by **flow charts**. *p137*

3 Algorithms can be written as a series of instructions written in **pseudo English**. *p140*

Test yourself

What to review

The algorithm below is used to find the highest mark, given as percentages, of six students taking a test.

```
LINE 10   Let M = 0
LINE 20   For I = 1 to 6
LINE 30   Input X
LINE 40   If X < M then goto LINE 60
LINE 50   Let M = X
LINE 60   Next K
LINE 70   Print M
LINE 80   End
```

	What to review
1 State the purpose of LINE 10.	*Section 8.3*
2 There is a mistake on LINE 60. Write down a correct LINE 60.	*Section 8.3*
3 Write the algorithm as a flow diagram.	*Section 8.2*
4 Trace the algorithm in the case where $X = 25, 73, 18, 82, 47, 93$.	*Section 8.3*
5 Write down an amended algorithm that would find the minimum of a set of six numbers.	*Section 8.3*
6 Write down an amended algorithm that could be used if the number of marks to be input was unknown.	*Section 8.3*

Test yourself ANSWERS

1 Max. mark = 0

2 Next I

3

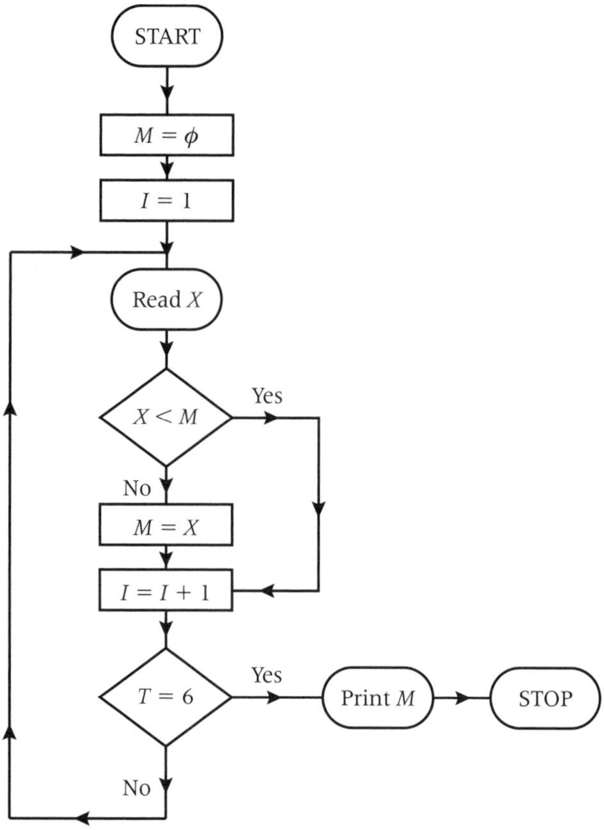

4

M	I	X
0	1	25
25	2	73
73	3	18
	4	82
82	5	47
	6	93
93		

5 LINE 10 Let $M = 100$
LINE 40 If $X > M$

6 Delete LINES 20 and 60 and add a new LINE 60 goto 30.
Add a rogue value of -1 to the end of the X values:
If $X = -1$ then goto LINE 70

8

CHAPTER 9

Linear programming

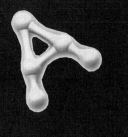

Learning objectives

After studying this chapter you should be able to:

- formulate linear programming problems from real-life problems
- identify feasible regions for linear programming problems
- form appropriate objective functions
- give graphical representations of feasible regions
- solve linear programming problems.

9.1 Introduction

The methods of linear programming were originally developed between 1945 and 1955 by American mathematicians to solve problems arising in industry and economic planning. Many such problems involve constraints on the size of the workforce, the quantities of raw materials available, the number of machines available and so on.

Problems occurring in industry have many variables and have to be solved by computer. For example, in oil refineries, problems arise with hundreds of variables and tens of thousands of constraints. Another application is in determining the best diet for farm animals such as pigs. In order to maximise profit a pig farmer needs to ensure that the pigs are fed appropriate food and sufficient quantities of it to produce lean meat. The pigs require a daily allocation of carbohydrate, protein, amino acids, minerals and vitamins. Each involves various components. For example, the mineral content includes calcium, phosphorus, salt, potassium, iron, magnesium, zinc, copper, manganese, iodine and selenium. All these dietary constituents should be present in the correct amounts. Linear programming has led to a computer program for use by farmers and companies producing animal feeds which enables them to provide the right diet for pigs at various stages of development, such as the weaning, growing and finishing stages. The program involves 20 variables and 10 equations!

Undoubtedly linear programming is one of the most widespread methods used to solve management and economic problems, and it has been applied in a wide variety of situations and contexts. In this chapter our problems usually only involve two variables, enabling us to solve them graphically. What is important is to understand the principles involved so that they can then be applied to more complicated situations. (In particular we return to this topic in the follow-up book D2.)

9.2 Graphs of inequalities

Inequalities in one variable x are usually easy to solve. For example, if you require both $x \geqslant 3$ and $x \leqslant 10$, then the solution is all x values from 3 up to 10 inclusive. However, inequalities in two variables are harder to digest. For example, if you were asked to solve the pair of inequalities:

$$x + y \geqslant 5 \text{ and } 2x + y \leqslant 12$$

then it is not immediately clear how to describe all those x and y values that work. Certainly $x = 4$, $y = 2$ is a solution, as is $x = 4\frac{1}{2}$, $y = 3$, whereas $x = 5$, $y = 3$ is not since the second inequality fails.

A question with two variables is easier to answer in a two-dimensional way, and so we use graphical methods. (Here we are reverting to the traditional meaning of a graph as something drawn on graph paper!) Recall that to draw a graph we usually have a horizontal x-axis and a vertical y-axis, and each point in the plane is determined by its x-coordinate and its y-coordinate. In the (x, y)-plane illustrated we have marked the point $(3, 2)$, with x-coordinate 3 and y-coordinate 2, and the point $(-4, -1)$, with coordinates -4 and -1.

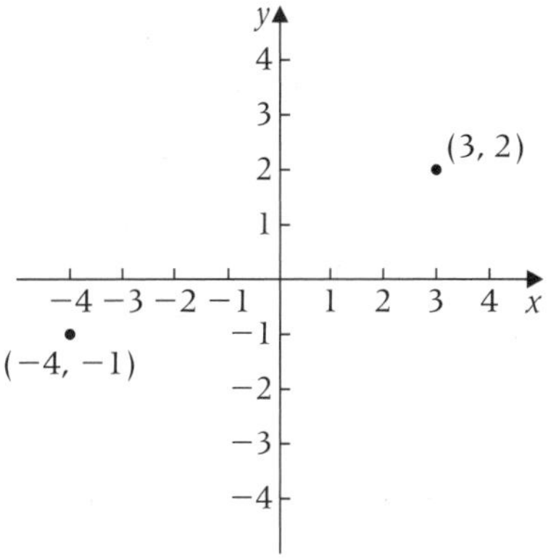

A linear equation in x and y is one like:

$$2x + 3y = 6$$
↖ ↑ ↗
numbers

The points in the (x, y)-plane that satisfy a linear equation form a straight line (hence the name) and the line $2x + 3y = 6$ is illustrated.

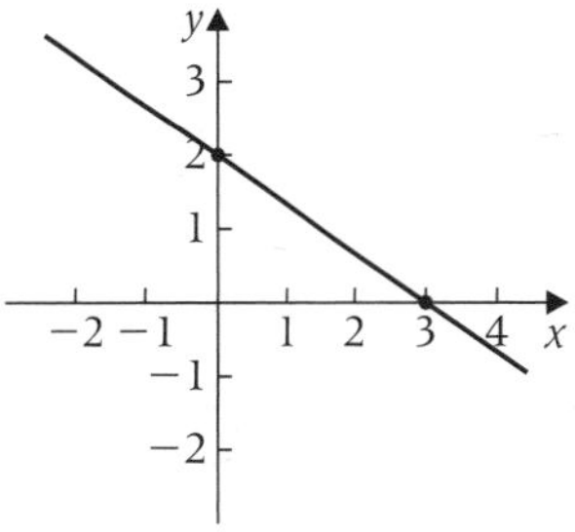

Having illustrated a line it is now straightforward to illustrate a linear **in**equality such as:

$$2x + 3y \leqslant 6.$$

In the above example it is all points on or below the line that are illustrated. So, for example, $(0, 0)$ is below the line and $x = 0$, $y = 0$ satisfies the inequality, whereas $(3, 1)$ is above the line and $x = 3$, $y = 1$ does not satisfy it.

9

Worked example 9.1

Find the region that satisfies $2x + y \geqslant 2$.

Solution

Start by drawing the line $2x + y = 2$. This will form the boundary of our required region. The inequality $2x + y \geqslant 2$ is satisfied by all points on one side of the line. To identify which side, you can test the point (0, 0) – this does not satisfy the inequality, so the region to the right of the line is the solution. The **excluded** region is on the **shaded** side of the line.

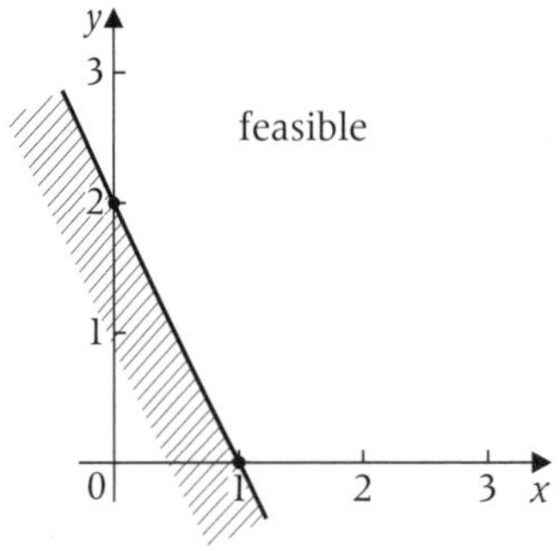

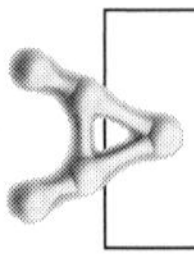

The answer is the unshaded side of the line: it is called the **feasible region**.

Incidentally, we have restricted the picture to the **positive quadrant** of the (x, y)-plane where the x- and y-coordinates are 0 or more. This is because linear programming problems invariably concern non-negative x and y.

Worked example 9.2

Find the region that satisfies $2x + y \geqslant 2$ and $x + 2y \geqslant 2$.

Solution

Here we are solving a pair of simultaneous inequalities. We have already solved the first inequality, and if you add on the graph of the second inequality, you obtain the region as shown in the diagram, remembering that shaded regions are excluded.

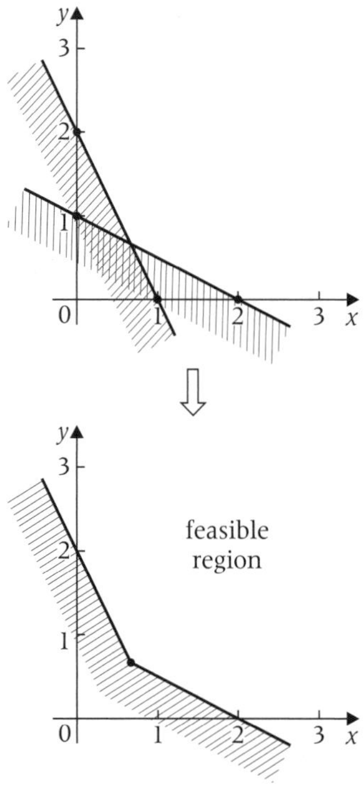

Combining the two inequalities gives the solution region as shown.

The **vertex** (or 'corner') of the solution region is the point where the two lines cross. This can be found from the picture or from solving the simultaneous equations

$$2x + y = 2$$

and $\quad x + 2y = 2$

to give $\quad x = y = 2/3$.

Worked example 9.3

Find the region that satisfies all of the following inequalities:

$$x + y \geqslant 2$$
$$3x + y \geqslant 3$$
$$x + 3y \geqslant 3$$

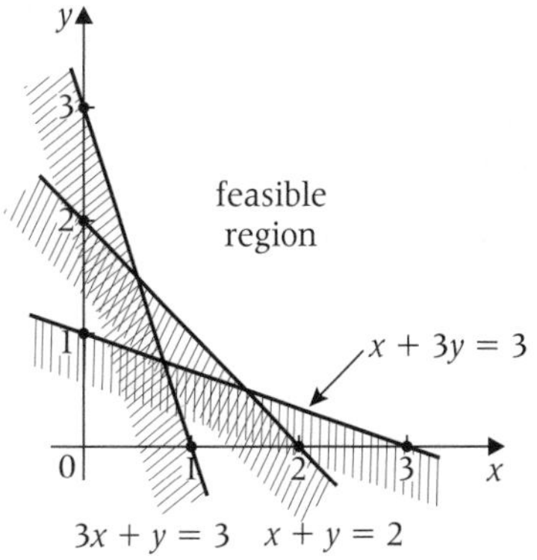

Solution

As before, the graph of the three inequalities is first drawn and the region in which **all** three are satisfied is noted.

Note that if you had wanted to solve:

$$x + y \leqslant 2$$
$$3x + y \geqslant 3$$
$$x + 3y \geqslant 3$$

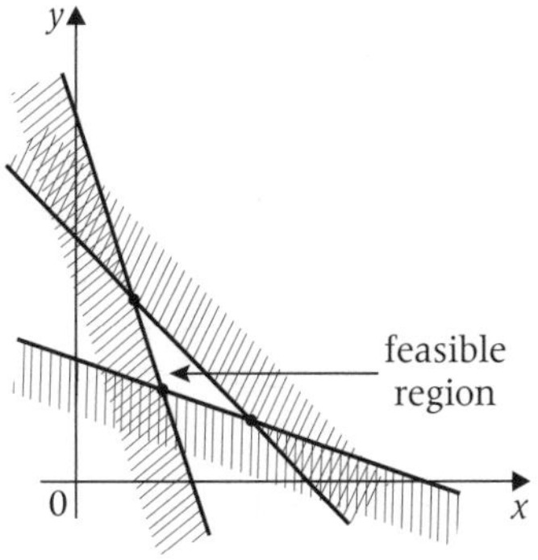

then the solution would have been the triangular region completely bounded by the three lines; in general the word **finite** is used for such bounded regions.

In the pictures of the regions we have highlighted the vertices: these will play a significant part in our work on linear programming.

EXERCISE 9A

1 Solve $y \geqslant 0$, $x + y \leqslant 2$ and $y - 2x \leqslant 2$.

2 Find the solution set for $x + y \leqslant 1$ and $3x + 2y \leqslant 6$.

3 Find the region that satisfies:

$$x + y \leqslant 2$$
$$x + 4y \leqslant 4$$
$$x \geqslant 0$$
$$y \geqslant 0$$

Is it finite?

9.3 Formulation of the problem

We are now in a position to use our knowledge of inequalities from the previous section to illustrate **linear programming** using the following case study concerning the manufacture of printed circuits.

Suppose a manufacturer of printed circuits has a stock of 200 resistors, 120 transistors and 150 capacitors, and is required to produce two types of circuits.

Type A requires 20 resistors, 10 transistors and 10 capacitors.

Type B requires 10 resistors, 20 transistors and 30 capacitors.

If the profit on type A circuits is £5 and that on type B circuits is £12, how many of each circuit should be produced in order to maximise the profit?

We will not actually solve this problem yet, but show how it can be formulated as a linear programming problem. There are three vital stages in the formulation, namely:

(a) What are the unknowns?
(b) What are the constraints?
(c) What is the profit/cost to be maximised/minimised?

For this problem:

(a) What are the unknowns?
Clearly the number of type A and type B circuits produced; so we define:

x = number of type A circuits produced
y = number of type B circuits produced

(b) What are the constraints?
There are constraints associated with the total number of resistors, transistors and capacitors available.

Resistors Since each type A requires 20 resistors and each type B requires 10 resistors, then, as there are only 200 available,

$$20x + 10y \leqslant 200 \text{ (or } 2x + y \leqslant 20)$$

Transistors Similarly, since each type A requires 10 transistors, each type B requires 20 and there are only 120 available, we have

$$10x + 20y \leqslant 120 \text{ (or } x + 2y \leqslant 12)$$

Capacitors Similarly:

$$10x + 30y \leqslant 150 \text{ (or } x + 3y \leqslant 15)$$

Finally you must state the obvious (but nevertheless important) inequalities

$$x \geqslant 0, y \geqslant 0$$

(c) What is the profit?
Since each type A gives £5 profit and each type B gives £12 profit, the total profit is £P, where:

$$P = 5x + 12y$$

This is called the **objective function**.

We can now summarise the problem as:

maximise $P = 5x + 12y$
subject to $2x + y \leqslant 20$
$x + 2y \leqslant 12$
$x + 3y \leqslant 15$
$x \geqslant 0$
$y \geqslant 0$

This is called a **linear** programming problem since both the objection function P and the constraints are all linear.

In this particular example you should be aware that x and y can only be integers since it is not sensible to consider fractions of a printed circuit. In all linear programming problems you need to consider if the variables are integers.

We shall actually solve this problem in the next section, but note for the moment that, for example, $x = 5$, $y = 3$ satisfies all the inequalities and gives a profit of $P = £61$. You might try to find other points that work and that increase P. We shall learn the proper method in the next section but at this stage, we will not continue with finding the actual solutions but concentrate on further practice in formulating problems of this type.

Worked example 9.4

A small firm builds two types of garden shed.

Type A requires 2 hours of machine time and 5 hours of craftsman time.

Type B requires 3 hours of machine time and 5 hours of craftsman time.

Each day there are 30 hours of machine time available and 60 hours of craftsman time. The profit on each type A shed is £60 and on each type B shed is £84.

Formulate the appropriate linear programming problem.

Solution

The key stage is the first one, namely that of identifying the unknowns. Once you have done this successfully, it should be straightforward to express both the constraints and the profit function in terms of the unknowns.

In this case the two things that we can choose are the number of sheds to be produced of each type, so these must form our two variables.

(a) Unknowns
Define:

x = number of type A sheds produced each day
y = number of type B sheds produced each day

(b) Constraints
Machine time: $2x + 3y \leqslant 30$
Craftsman time: $5x + 5y \leqslant 60$ (or $x + y \leqslant 12$)
and $x \geqslant 0, y \geqslant 0$

(c) Profit
$P = 60x + 84y$

In summary, the linear programming problem is

$$\begin{aligned} \text{maximise} \quad & P = 60x + 84y \\ \text{subject to} \quad & 2x + 3y \leqslant 30 \\ & x + y \leqslant 12 \\ & x \geqslant 0 \\ & y \geqslant 0 \end{aligned}$$

EXERCISE 9B

[These exercises concern formulating linear programming problems. They are solved in Exercise 9C. Generally examination questions will consist of both the formulation and the solution.]

1 A tourist decides to buy some beer and wine in a French supermarket. He has to decide how many crates of each to buy.

There is room in the boot of his car for up to four crates.

A crate of beer weighs 10 kg and a crate of wine weighs 20 kg. The maximum extra weight that the car will take is 60 kg.

(a) Formulate the given conditions as four inequalities in two variables of your own choice.

(b) If each crate of beer saves the tourist £10 and each crate of wine saves him £15, what is the objective function and should it be maximised or minimised?

2 Ann and Margaret run a small business in which they work together making blouses and skirts. Each blouse takes 1 hour of Ann's time together with 1 hour of Margaret's time. Each skirt involves Ann for 1 hour and Margaret for half an hour. Ann has 7 hours available each day and Margaret has 5 hours each day.

They could just make blouses or they could just make skirts or they could make some of each.

Their first thought was to make the same number of each, but they get £8 profit on a blouse and only £6 on a skirt.

(a) Formulate the problem as a linear programming problem.

(b) Find three solutions that satisfy the constraints.

3 A distribution firm has to transport 1200 packages using large vans which can take 200 packages each and small vans which can take 80 packages each. The cost of running each large van is £40 and of each small van is £20. Not more than £300 is to be spent on the job. The number of large vans must not exceed the number of small vans.

Formulate this problem as a linear programming problem given that the objective is to **minimise** costs. [A]

4 A firm manufacturers wood screws and metal screws. All the screws have to pass through a threading machine and a slotting machine. A box of wood screws requires 3 minutes on the slotting machine and 2 minutes on the threading machine. A box of metal screws requires 2 minutes on the slotting machine and 8 minutes on the threading machine. In a week, each machine is available for 60 hours.

There is a profit of £10 per box on wood screws and £17 per box on metal screws.

Formulate this problem as a linear programming problem given that the objective is to **maximise** profit. [A]

5 A factory employs unskilled workers earning £135 per week and skilled workers earning £270 per week. It is required to keep the weekly wage bill to £24 300 or less.

The machines require a minimum of 110 operators, of whom at least 40 must be skilled. Union regulations require that the number of skilled workers should be at least half the number of unskilled workers.

If x is the number of unskilled workers and y the number of skilled workers, write down all the constraints to be satisfied by x and y.

9.4 Graphical solutions

In the previous section we worked through problems that led to a linear programming problem in which a **linear** function of x and y is to be maximised (or minimised) subject to a number of **linear** inequalities to be satisfied.

Fortunately problems of this type with just two variables can easily be solved using a graphical method. The method will first be illustrated using the example concerning the manufacture of circuits, from the beginning of section 4.3. It resulted in the linear programming problem:

$$\begin{aligned}
\text{maximise} \quad & P = 5x + 12y \\
\text{subject to} \quad & 2x + y \leqslant 20 \\
& x + 2y \leqslant 12 \\
& x + 3y \leqslant 15 \\
& x \geqslant 0 \\
& y \geqslant 0
\end{aligned}$$

We can illustrate the **feasible** (i.e. allowable) region by drawing graphs of all the inequalities, as we did in section 9.3, shading out the regions not allowed. This is illustrated in the figure below.

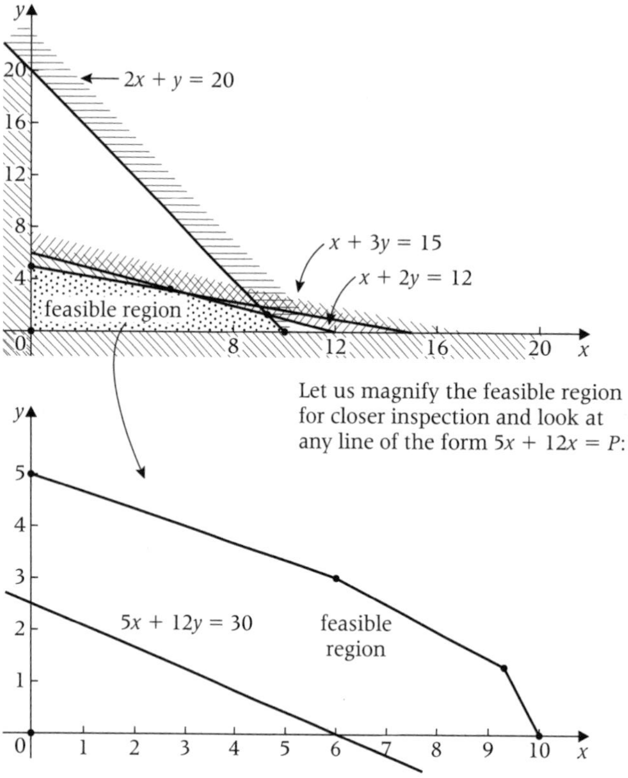

We have drawn the line

$$5x + 12y = 30$$

as an example. Any point (x, y) on that line and in the feasible region will give us an x and y satisfying all the inequalities and making a profit of £30.

How can that £30 be increased? By moving the line, keeping it parallel. For example, the lines:

$$5x + 12y = 60 \quad \text{and} \quad 5x + 12y = 66$$

are shown as on the right:

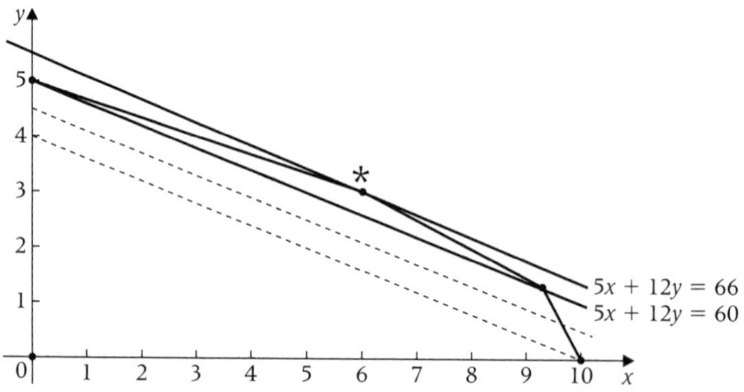

You can see that if you move the line any further it will no longer meet the feasible region. So we conclude that the maximum profit is £66 and this is achieved at the point marked *. You can either read off the coordinates of * from your graph (provided that it was drawn

very accurately) or work out the coordinates of * by solving the simultaneous equations of the two lines meeting there, namely:

$x + 3y = 15$ and $x + 2y = 12$

This gives $x = 6$ and $y = 3$, therefore the maximum profit of £66 is achieved when $x = 6$ and $y = 3$ (i.e. when six type A and three type B circuits are made).

> In practice you do not have to draw lots of profit lines. You can draw any one line and then move your ruler parallel to it to see where it 'just' leaves the region. The point * is known as the **optimal point**, and the value of the profit there is the **optimal value**.

Worked example 9.5

A small firm builds two types of garden shed.

Type A requires 2 hours of machine time and 5 hours of craftsman time.

Type B requires 3 hours of machine time and 5 hours of craftsman time.

Each day there are 30 hours of machine time available and 60 hours of craftsman time. The profit on each type A shed is £60 and on each type B shed is £84.

How many of each type should the firm make each day in order to maximise its profits?

Solution

This is Worked example 9.4 revisited. In the earlier example we formulated the problem and got:

$$\begin{aligned} \text{maximise} \quad & P = 60x + 84y \\ \text{subject to} \quad & 2x + 3y \leqslant 30 \\ & x + y \leqslant 12 \\ & x \geqslant 0 \\ & y \geqslant 0 \end{aligned}$$

where x = number of type A produced each day and y = number of type B produced each day.

The next stage is to illustrate the feasible region, which turns out to be the finite region shown. Maximising problems invariably lead to finite feasible regions.

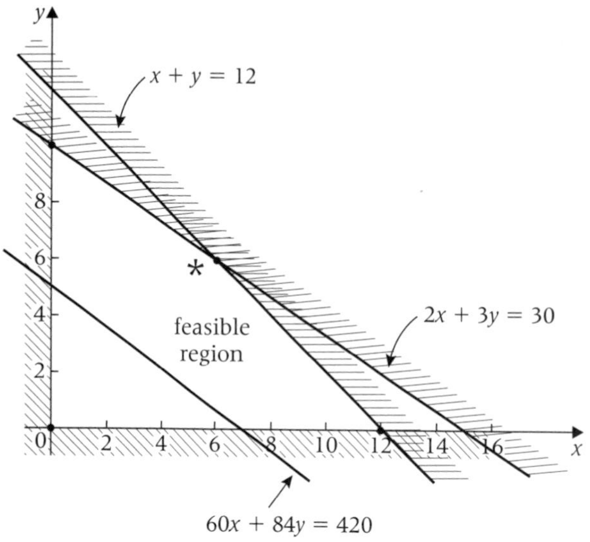

Now we need to consider the profit:

$$P = 60x + 84y$$

The line $60x + 84y = 420$ has been added to the picture. All 'profit lines' will be parallel to this and, the further up the picture they move, the greater the profit will be. You can see by sliding your ruler up the paper parallel to that profit line that you will touch the feasible region last of all at the vertex *.

The coordinates of *, found by solving:

$$x + y = 12 \text{ and } 2x + 3y = 30$$

are $x = 6$, $y = 6$ and they give:

$$P = 60x + 84y = 864$$

The profit is therefore maximised at £864 when the firm builds six type A and six type B sheds each day, and the problem is solved.

There is an alternative method worth mentioning. You will have seen from the way that we 'slide' the profit line across the feasible region that the maximum is bound to be reached at one of the vertices. Rather than considering profit lines you could instead work out the coordinates of all the vertices, calculate the profit at each of them and choose the largest.

In the previous example the feasible region had four vertices: (0, 0), (0, 10), (6, 6) and (12, 0). The profits there would be

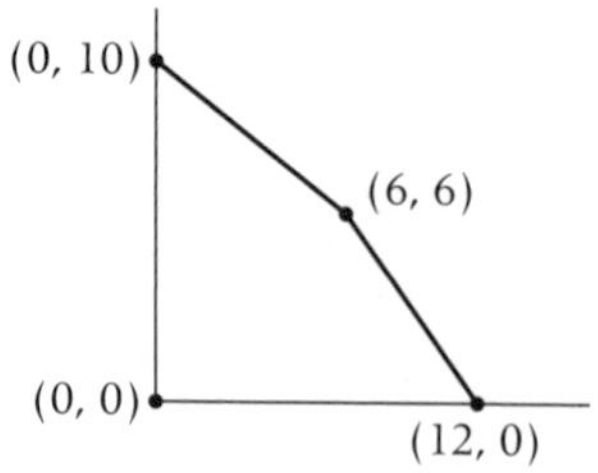

(0, 0): $P = 60x + 84y = 0$
(0, 10): $P = 60x + 84y = 840$
(6, 6): $P = 60x + 84y = \mathbf{864}$
(12, 0): $P = 60x + 84y = 720$

which confirms that the maximum of £864 is reached when $x = 6$ and $y = 6$.

Worked exam question 9.1

A farmer has up to 20 hectares for growing barley and swedes. The farmer has to decide how much of each to grow. The cost per hectare for barley is £30 and for swedes is £20. The farmer has budgeted a maximum of £480 for these costs.

Barley requires 1 effort-day per hectare and swedes require 2 effort-days per hectare. There are 36 effort-days available.

The profit on barley is £100 per hectare and on swedes is £120 per hectare.

Find the number of hectares of each crop the farmer should sow to maximise profits.

Solution

The problem is formulated as a linear programming problem:

(a) Unknowns
x = number of hectares of barley
y = number of hectares of swedes

(b) Constraints
Land: $x + y \leqslant 20$
Cost: $30x + 20y \leqslant 480$ (or $3x + 2y \leqslant 48$)
Manpower: $x + 2y \leqslant 36$

(c) Profit
$P = 100x + 120y$

To summarise:

$$\begin{aligned} \text{maximise} \quad & P = 100x + 120y \\ \text{subject to} \quad & x + y \leqslant 20 \\ & 3x + 2y \leqslant 48 \\ & x + 2y \leqslant 36 \\ & x \geqslant 0 \\ & y \geqslant 0 \end{aligned}$$

The feasible region is identified by the finite region enclosed by the five inequalities, as shown below. Of course, two of the boundaries are just the axes.

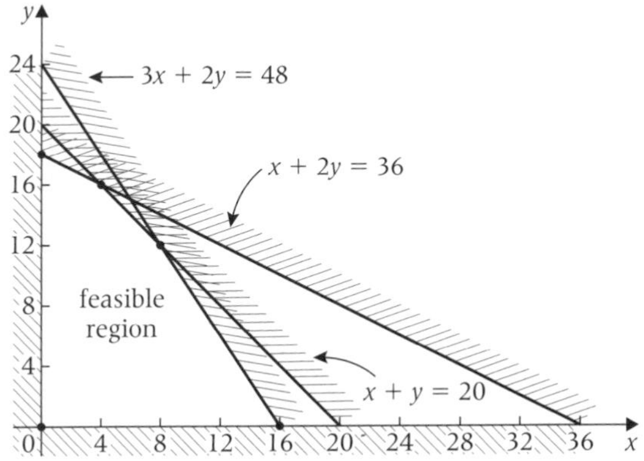

Let us look at the feasible region more clearly:

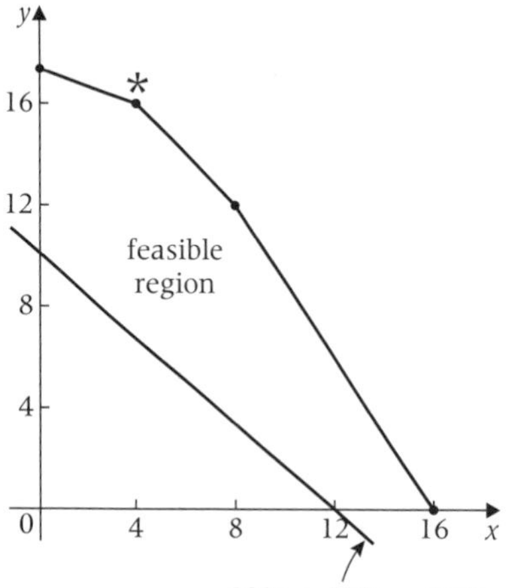

There are now alternative ways of finding the maximum profit $P = 100x + 120y$. One is to draw a typical profit line, such as $100x + 120y = 1200$, as shown.

Sliding your ruler parallel to that line and moving it up the page shows that the last point at which you touch the feasible region is the vertex *. This is the intersection of the lines:

$x + 2y = 36$ and $x + y = 20$

which can be solved to give $x = 4$, $y = 16$,

hence the maximum profit of:

$$P = 100x + 120y = 100 \times 4 + 120 \times 16 = 2320$$

is achieved by growing 4 hectares of barley and 16 hectares of swedes.

The alternative approach is to find the coordinates of all the vertices and calculate the profit at each, since we know that the maximum will be achieved at a vertex of the feasible region. Its five vertices are (0, 0), (0, 18), (4, 16) (as above), (8, 12) (the intersection of $3x + 2y = 48$ and $x + y = 20$) and (16, 0). The profits there would be

(0, 0): $P = 100x + 120y = 0$
(0, 18): $P = 100x + 120y = 2160$
(4, 16): $P = 100x + 120y =$ **2320**
(8, 12): $P = 100x + 120y = 2240$
(16, 0): $P = 100x + 120y = 1600$

and so (as expected) the maximum is £2320 and happens when $x = 4$ and $y = 16$.

So far our worked examples have all concerned maximising profits but there is another class of linear programming problems that concern minimising costs. Luckily the method of solution is very similar.

Worked exam question 9.2

Pigs have certain dietary requirements of carbohydrates, proteins and minerals. A farmer can provide these by buying type X or type Y pig food mix in bags. The contents of the bags and the requirements of the farmer's pigs are shown in the table below.

	Carbohydrates	Proteins	Minerals
Units in each type X bag	5	5	2
Units in each type Y bag	15	2	2
Daily requirements for pigs	30	10	8

If each type X bag costs £8 and each type Y bag costs £16, how many bags of each should the farmer buy to satisfy the pigs' dietary requirements at the minimum cost?

Solution

(a) Unknowns
x = number of type X bags
y = number of type Y bags

(b) Constraints
Carbohydrates: $5x + 15y \geqslant 30$ (or $x + 3y \geqslant 6$)
Proteins: $5x + 2y \geqslant 10$
Minerals: $2x + 2y \geqslant 8$ (or $x + y \geqslant 4$)

(c) Cost
$C = 8x + 16y$

The linear programming problem is therefore:

$$\begin{aligned} \text{minimise} \quad & C = 8x + 16y \\ \text{subject to} \quad & x + 3y \geqslant 6 \\ & 5x + 2y \geqslant 10 \\ & x + y \geqslant 4 \\ & x \geqslant 0 \\ & y \geqslant 0 \end{aligned}$$

The feasible region is as shown. Typically for minimising problems it is not finite.

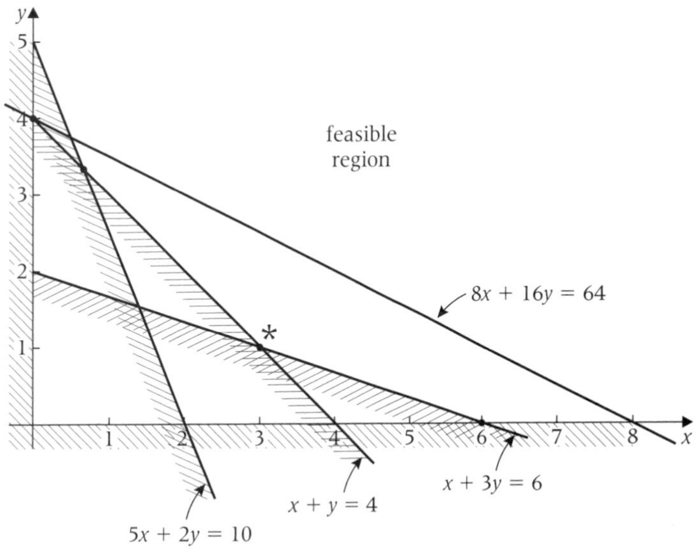

We have drawn one of the cost lines, $8x + 16y = 64$. (*Examiner's tip*: there is no need to clutter your picture with lots of profit lines.) This time we wish to minimise the cost and so we slide the parallel line downwards and the last point at which it touches the region is the vertex *. This is the intersection of

$x + y = 4$ and $x + 3y = 6$

which gives $x = 3$ and $y = 1$. Hence the dietary requirements of the pigs are satisfied at minimum cost if the farmer buys three bags of type X and one bag of type Y. The minimum cost is then

$$C = 8x + 16y = 8 \times 3 + 16 \times 1 = 40$$

i.e. £40.

The alternative approach, with vertices rather than cost lines, is to calculate the coordinates of the four vertices of the feasible region and the cost at each:

$(0, 5)$: $C = 8x + 16y = 80$
$(\frac{2}{3}, 3\frac{1}{3})$: $C = 8x + 16y = 58\frac{2}{3}$
$(3, 1)$: $C = 8x + 16y = \mathbf{40}$
$(6, 0)$: $C = 8x + 16y = 48$

confirming that the minimum of £40 happens when $x = 3$ and $y = 1$.

> Had the minimum occurred at $(\frac{2}{3}, 3\frac{1}{3})$ we would have considered integer points nearby and in the feasible region; $(1, 3)$ for example.

Worked example 9.6

Solve the linear programming problem:

$$2x + y \leqslant 16$$
$$x + 3y \leqslant 20$$
$$x \geqslant 0, y \geqslant 0$$

(a) Maximise $x + y$.

(b) Given that x and y are integers, maximise:

(i) $x + y$

(ii) $x + 4y$

(iii) $5x + 4y$

(iv) $4x + 5y$

Solution

The feasible region is identified by the feasible region as shown.

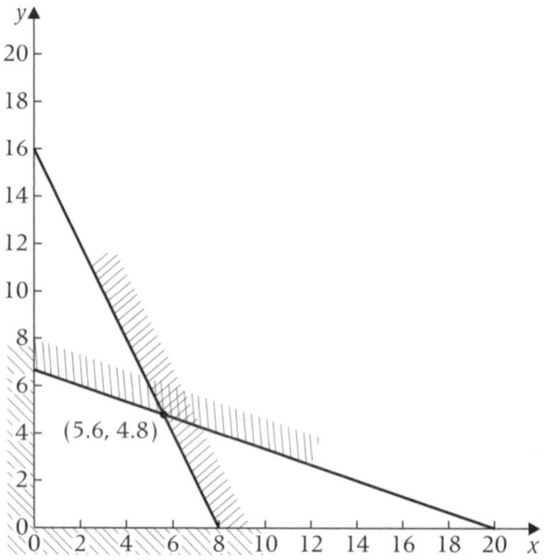

(a) The maximum value of $x + y$ occurs at the intersection of the two lines, i.e. the point with coordinates $(5.6, 4.8)$.

The maximum value of $x + y = 10.4$.

(b) As x and y have to be integers, the solution cannot be the same as the solution to **(a)**.

We must consider integer values within the feasible region.

It cannot be assumed that by rounding down the exact solution we obtain the best integer solution.

If we redraw the feasible region and highlight integer points within this region, we obtain the diagram below.

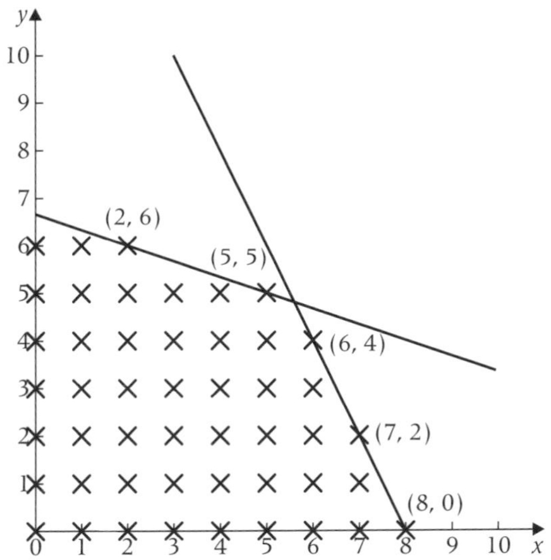

It is not accurate enough to draw our objective line to find the point farthest from the origin, we must consider possible integer solutions.

The possible maximising solution will occur at one of the following points:

(2, 6), (5, 5), (6, 4), (7, 2), (8, 0)

By substituting these values into the respective maximising function we can solve the remaining parts of the question.

(i) Maximum value of $x + 4y$ occurs at (2, 6) and has a value of 26.

(ii) Maximum value of $5x + 4y$ occurs at (6, 4) and has a value of 46.

(iii) Maximum value of $4x + 5y$ occurs at (5, 5) and has a value of 45.

Problems involving three variables

All the examples in this chapter have involved two variables so that we can use an (x, y)-plane to illustrate them and then use our graphical method. If a corresponding problem involved three variables then our method would require a three-dimensional model, which is not very practical. More than three variables could not be modelled in this way. An algebraic alternative is therefore needed that can cope with any number of variables. The simplex method does just that and it forms part of the follow-up course D2.

EXERCISE 9C

1–4 For each of the questions 1–4 of Exercise 9B, in which a linear programming problem was formulated, use a graphical method to solve the problem.

5 Illustrate the region satisfying the inequalities found in question 5 of Exercise 9B.

(a) What is the maximum number of skilled workers possible?

(b) What is the maximum number of unskilled workers possible?

6 **(a)** Illustrate the region satisfying:

$$x + 2y \leqslant 4$$
$$3x + y \leqslant 6$$
$$x \geqslant 0$$
$$y \geqslant 0$$

(b) At which points in the region are the following functions maximised?

$$P = 5x + y$$
$$Q = x + y$$
$$R = x + 3y$$

7 A firm manufactures two types of box, each requiring the same amount of material.

They both go through a folding machine and a stapling machine.

Type A boxes require 4 seconds on the folding machine and 3 seconds on the stapling machine.

Type B boxes require 2 seconds on the folding machine and 7 seconds on the stapling machine.

Each machine is available for 1 hour.

There is a profit of 40p on type A boxes and 30p on type B boxes.

How many of each type should be made to maximise the profit? [A]

MIXED EXERCISE

1 The Elves toy company makes toy trains and dolls' prams, which use the same wheels and logo stickers.

Each train requires 8 wheels and 2 logo stickers.

Each pram requires 8 wheels and 3 logo stickers.

The company has 7200 wheels and 2200 logo stickers available.

The company is to make at least 300 of each type of toy and at least 800 toys in total.

The company sells each train for £20 and each pram for £25.

The company makes and sells x trains and y prams.

The company needs to find its minimum and maximum total income, £T.

(a) Formulate the company's situation as a linear programming problem.

(b) Draw a suitable diagram to enable the program to be solved graphically, indicating the feasible region and the direction of the objective line.

(c) Use your diagram to find the company's minimum and maximum total income, £T. [A]

2 The following graph shows the feasible region of a linear programming problem.

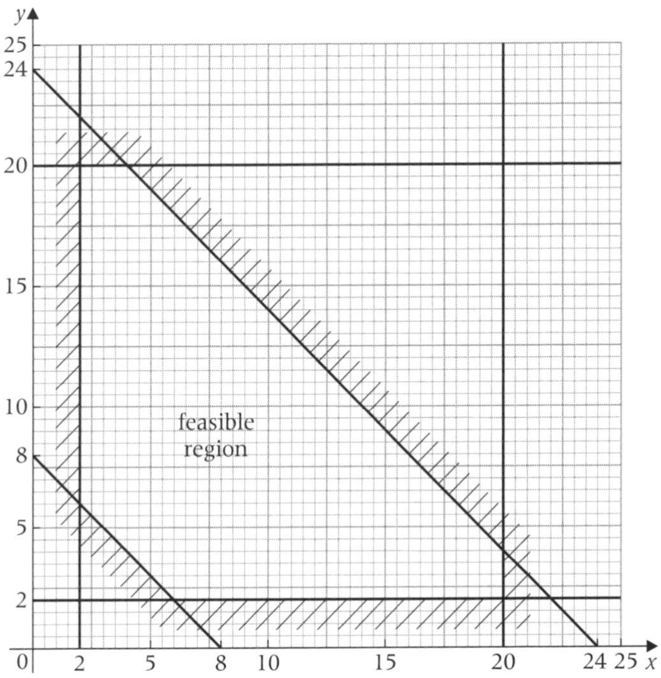

(a) Find the six inequalities that define the feasible region.

(b) On this feasible region, find:

(i) the maximum value of $x + 3y$,

(ii) the minimum value of $-x + 2y$. [A]

3 Every Saturday, David irons for the family. He irons three different types of clothing: dresses, shirts and jackets.

David knows that each Saturday:

he must iron at least 2 dresses, at least 5 shirts and at least 2 jackets;
he can only iron at most 28 items of clothing;
he takes 3 minutes to iron a dress, 2 minutes to iron a shirt and 2 minutes to iron a jacket.
The iron must be used for at least 40 minutes but not more than 60 minutes.

Each Saturday, David irons x dresses, y shirts and z jackets.

(a) Find six inequalities that model David's situation.

(b) On a particular Saturday, David irons the same number of jackets as dresses.

David's aim is to iron the maximum number of items of clothing.

(i) Show that the inequalities found in **(a)** simplify to

$$x \geqslant 2,\ y \geqslant 5,\ 2x + y \leqslant 28,\ 40 \leqslant 5x + 2y \leqslant 60.$$

(ii) Draw a suitable diagram to enable the problem to be solved graphically, indicating the feasible region and the direction of the objective line.

(iii) Use your diagram to find how many of each type of clothing David might iron to achieve his aim. [A]

4 A football team is performing badly. The manager of the team can buy some new players to try and gain some extra points.

A very simple model of the manager's situation is as follows.

He can buy new attackers for £3 million each and new defenders for £1 million each.

Each attacker will help the club to get two more points and each defender will help the club to get one more point.

The manager has £12 million to spend.

The team must gain at least 5 more points.

The manager must buy at least one defender and at least one attacker.

He must spend at least £6 million.

The manager buys x attackers and y defenders.

(a) Express the manager's situation in terms of inequalities suitable for investigating the problem by linear programming.

(b) Draw a suitable diagram to enable the problem to be investigated graphically, indicating the feasible region.

(c) Use your diagram to find:

(i) the maximum number of points the team can gain,

(ii) the numbers of attackers and defenders which will allow the team to gain the maximum number of points if the manager spends exactly £11 million. [A]

5 The following graph shows the feasible region of a linear programming problem.

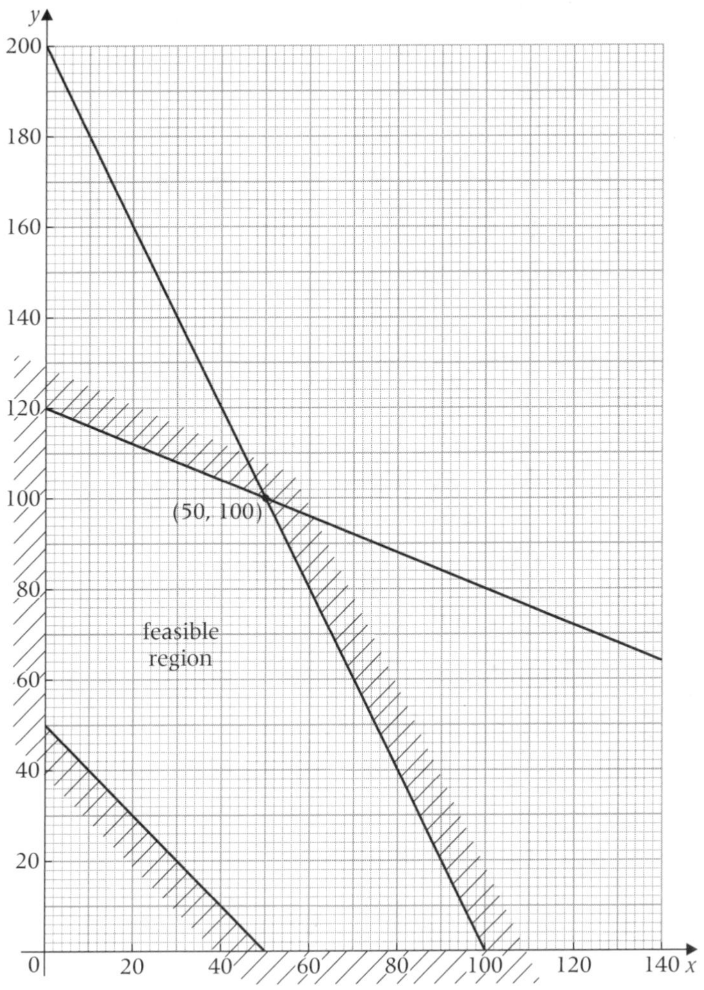

(a) On this feasible region find:

(i) the maximum value of the function $2x + 3y$,

(ii) the minimum value of the function $4x + y$.

(b) Find the **five** inequalities that define the feasible region. [A]

6 A garage owner is to restock with new cars of three different types: hatchbacks, sports cars and estate cars.

She decides that:

the number of hatchbacks must be greater than the combined total number of sports cars and estates,
the number of sports cars must not exceed 15% of the total number of cars,
there must be at least as many sports cars as estates,
the number of estates must be less than 60% of the number of hatchbacks.

Each hatchback costs £8000, each sports car costs £12 000, and each estate costs £14 000.

The garage owner buys x hatchbacks, y sports cars and z estates.

She wishes to minimise her total cost, £C.

(a) Formulate the above situation as a linear programming problem, simplifying your inequalities so that all the coefficients are integers.

(b) The garage owner buys 6 sports cars. By using a graphical method, find her minimum total cost in this case. [A]

7 A linear programming problem is represented by the following:

$x + y \geqslant 20$
$y \leqslant 2x$
$x + y \leqslant 60$
$x + 2y \leqslant 80$
$x \geqslant 0$
maximise $F = 2x + 5y$

(a) Draw a suitable diagram to enable the problem to be solved graphically, indicating the feasible region and the direction of the objective line.

(b) Use your diagram to find the maximum value of F on the feasible region. [A]

8 Stephanie works at a garden centre. Each day during May she has to make hanging baskets using three types of plant: ivy (I), lobelia (L) and primula (P).

She makes three different types of hanging basket: standard, superior and luxury.

Standard baskets contain 1 ivy, 2 lobelia and 3 primula plants.
Superior baskets contain 2 ivy, 3 lobelia and 3 primula plants.
Luxury baskets contain 3 ivy, 5 lobelia and 4 primula plants.

Each day Stephanie must use **at least** 30 ivy, 50 lobelia and 40 primula plants, but **not** more than 200 plants in total, and **at least** 40% of the plants she uses must be lobelia.

In a day Stephanie makes x standard baskets, y superior baskets and z luxury baskets. Find **five** inequalities in x, y and z which must be satisfied, simplifying each inequality where possible. [A]

9 A knitwear company employs machinists to make cardigans and jumpers on knitting machines.

In a week a machinist makes x cardigans and y jumpers.

Each machinist must make at least 5 cardigans and at least 20 jumpers each week.

Each cardigan uses 3 ounces of wool and has 6 buttons.

Each jumper uses 3.5 ounces of wool and has 2 buttons.

In any one week a machinist has only 210 ounces of wool and 180 buttons available.

A machine can knit a cardigan in 30 minutes and a jumper in 40 minutes. To ensure efficiency, the company requires each machinist to use a machine for at least 20 hours each week.

(a) Find **five** inequalities that model the machinist's situation.

(b) Draw a suitable diagram to represent this problem graphically, indicating the feasible region.

(c) The machinist is paid £3.00 for each cardigan and £2.50 for each jumper that he makes.

  (i) Draw an objective line that will represent his pay, £P, for the week.

  (ii) Find values of x and y that will maximise his weekly pay and calculate this pay.

(d) The pay structure is to be altered so that he will be paid £2.50 for each cardigan and £4.00 for each jumper. Determine his new maximum pay. [A]

10 The Carryit Company has 1000 large packing cases to transport from Liverpool to Southampton; x of them by road, y of them by rail, and the rest by sea. The company intends to transport at least 10% and at most 50% of the cases by sea. Furthermore, the number transported by sea must be at most double the number transported by road.

(a) Write down, in terms of x and y, the number of cases transported by sea. Hence show that x and y must satisfy the inequalities:

$$500 \leqslant x + y \leqslant 900 \quad \text{and} \quad 3x + y \geqslant 1000$$

(b) Illustrate on graph paper the region of those (x, y) which satisfy $x \geqslant 0$ and $y \geqslant 0$ together with the inequalities in **(a)**.

(c) The cost of transporting the cases is

£50 for each case by road;
£55 for each case by rail; and
£25 for each case by sea.

  (i) Show that the total cost £T of sending the 1000 cases is given by

$$T = 25x + 30y + 25\,000$$

  (ii) Hence use **(b)** to find the minimum cost of transporting all the cases. State how many of the cases should be sent by each method in order to achieve that minimum.

9

(iii) In order to be more competitive the rail operator wants to reduce the cost of transporting each case to £R, where R is a whole number. The costs of transporting by road and sea will remain the same.

The rail operator chooses R as large as possible so that when minimising its transport costs the Carryit Company will have to send most of the 1000 cases by rail. Calculate the value of R. [A]

11 [One sheet of 2 mm paper is provided for use in answering this question.]

An insurance saleswoman can sell two types of policies, pension policies and life assurance policies.

Head-office require her to sell at least three policies of each type per week. Regulations stipulate that a pension policy needs 60 minutes of explanation but a life assurance policy needs 20 minutes of explanation.

The saleswoman knows that she can sell policies only in an evening and that in a normal week she has 15 hours available to sell policies.

Each pension policy has an annual premium income (API) of £600 whereas each life assurance policy has an API of £120.

Head-office requires a salesperson to sell, in a week, policies with a total API of at least £4800.

The saleswoman is paid a commission on each policy she sells. For each pension policy she is paid £48 and for each life assurance policy she is paid £36.

In a week the saleswoman sells x pension policies and y life assurance policies.

(a) **(i)** Show that $3x + y \leqslant 45$.

(ii) Find **three** further inequalities in x and y that model the saleswoman's situation.

(b) **(i)** Draw a suitable diagram to represent this problem graphically, indicating the feasible region.

(ii) Draw the line that will represent a weekly commission of £576. Hence find the vertex that will maximise her weekly commission.

(c) **(i)** State the values of x and y that maximise her commission.

(ii) If the saleswoman is paid this commission each week throughout a 45 week working year, calculate her income for the year.

(d) During a promotional week head-office changes her commission structure so that a pension policy pays £90 and a life assurance policy pays £18. All other constraints remain the same.

(i) She was paid £1170 in commission for the week. Determine **one** possible pair of values of x and y.

(ii) Calculate her maximum possible commission for the week. [A]

12 [One sheet of 2 mm graph paper is provided for use in answering this question.]

A businessman is considering starting a car hire company using Alpha cars which cost £12 500 each and Beta cars which cost £7500 each.

Research suggests that demand is such that he will need at least ten cars, but he has only £187 500 to buy cars.

Annual insurance and service costs are estimated at £1000 for each Alpha car and £2000 for each Beta car, and he wishes to limit his total annual expenditure on these items to £29 000.

Assume that he buys x Alpha and y Beta cars.

(a) Show that the above information can be modelled by the following inequalities.

$$x + y \geqslant 10 \quad 5x + 3y \leqslant 75 \quad x + 2y \leqslant 29$$

(b) The businessman estimates that he will make an annual profit of £1200 per car on both Alpha and Beta cars.

(i) Draw a suitable diagram to represent this problem graphically, indicating the feasible region.

(ii) Draw an objective line that will represent his annual profit, £P. Hence indicate the vertex that will correspond to the maximum value of P.

(iii) Write down the values of x and y which maximise P, and calculate the maximum value. [A]

13 [One sheet of 2 mm graph paper is provided for use in answering this question.]

A company is making two types of door, standard and luxury. Both types of door require the use of two different machines A and B.

Both types of door require 90 minutes on machine A. A standard door requires 60 minutes on machine B but a luxury door requires 120 minutes on this machine.
During any one week machine A can be used for a maximum of 20 hours and machine B can be used for a maximum of 25 hours.

The company makes a profit of £10 on each standard door and £12 on each luxury door. In a week the company makes x standard doors and y luxury doors.

(a) Show that the above information can be modelled by the following inequalities:

$$x \geqslant 0,\ y \geqslant 0,\ 3x + 3y \leqslant 40,\ x + 2y \leqslant 25$$

(b) **(i)** Draw a suitable diagram to represent the problem graphically, indicating the feasible region.

(ii) Draw an objective line that will represent the company's profit for the week, £P. Hence indicate the vertex, V, that could correspond to the maximum value of P.

(iii) State why this maximum value of P cannot be achieved.

(iv) Find values of x and y that will maximise the company's profit and calculate this profit. [A]

14 [One sheet of 2 mm graph paper is provided for use in answering this question.]

A factory manufactures three items: screws, nuts and bolts.

The items are first produced on a machine which is available for 4 hours per day. The machine takes 6 minutes to produce a screw, 4 minutes to produce a nut and 2 minutes to product a bolt. The items are then cleaned on a machine that is available for 55 minutes per day. Each screw takes 30 seconds to clean, each nut takes 40 seconds to clean and each bolt takes 60 seconds to clean.

(a) An apprentice at the factory produces x screws, y nuts and z bolts in a day. Find and simplify **two** inequalities each involving x, y and z, that model the conditions given above.

(b) As a further requirement the apprentice must produce the same number of bolts as nuts each day.

(i) Show that, with this additional constraint, the two inequalities found in **(a)** become:

$$x + y \leqslant 40 \quad \text{and} \quad 3x + 10y \leqslant 330$$

(ii) Draw a suitable diagram to represent this problem graphically, indicating the feasible region.

(iii) The apprentice has to make the largest possible total number of items each day. Draw an objective line that will represent his total daily output, T. Hence indicate the vertex that will correspond to the maximum value of T.

(iv) Calculate the maximum value of T.

(v) On a particular day the cleaning machine is available for only 48 minutes. Find the maximum number of items that the apprentice can make during this particular day. [A]

15 A building company has acquired a building site on which to build residential dwellings. It has permission to build at least 12 but not more than 16 dwellings. The company can build a combination of houses and bungalows on the site but it must build at least 4 of each.

Each house needs a plot of size 450 m^2 and will have a floor area of 200 m^2.
Each bungalow needs a plot of size 600 m^2 and will have a floor area of 150 m^2.
The cost of a plot is £50 per m^2 and the building costs are £100 per m^2 of floor area. The total value of land which can be used for building plots must not exceed £450 000. The total building costs must not exceed £300 000.

The company builds x houses and y bungalows.

(a) Show that the company's situation can be modelled by the following inequalities:

$12 \leqslant x + y \leqslant 16$, $x \geqslant 4$, $y \geqslant 4$, $3x + 4y \leqslant 60$,
$4x + 3y \leqslant 60$

(b) Draw a suitable diagram to represent this problem graphically, indicating the feasible region.

(c) The company sells all the houses and bungalows at a profit of £10 000 per dwelling. Find the **minimum** profit that the company is sure to make on this building site.

List **all** the different pairs of values of x and y that would produce this minimum profit.

(d) An alternative pricing structure is proposed in which houses and bungalows are all sold at the **same** price of £S. The company **maximises** the profit and this total profit is £190 000.

Find the largest and smallest possible values of S. [A]

Key point summary

Given a linear programming problem, the stages in its solution are as follows:

1 Illustrate the **feasible region** in an (x, y)-plane. *p152*

2 Choose your variables, express the constraints as **linear inequalities** and decide on an **objective function**. *p154*

3 Draw graphs of all the inequalities and find the highest (or lowest) that just touches the feasible region. *p158*

4 Find the x and y values of the **optimal point**. *p159*

5 Does the problem require x and y to be integers? If it does but your answer is not in integers, search for an integer solution nearby and in the feasible region. *p164*

Test yourself	What to review
To test yourself apply the stages to this very simple linear programming problem. A taxi driver has to decide how many adults and how many children to carry. He can take up to six people but that can include at most four adults. He charges £2 for each adult and £1 for each child.	
1 What are the variables? What inequalities do they satisfy? What is the objective function?	*Section 9.3*
2 Illustrate the inequalities and mark the feasible region.	*Sections 9.2 and 9.4*
3 Draw a profit line and find the highest point that just touches the feasible region.	*Section 9.4*
4 How can the taxi driver maximise his income?	*Section 9.4*

Test yourself ANSWERS

1 x = number of adults
y = number of children
$x + y \leqslant 6$
$x \leqslant 4$
(and $x \geqslant 0, y \geqslant 0$)
$P = 2x + y$

2 and 3

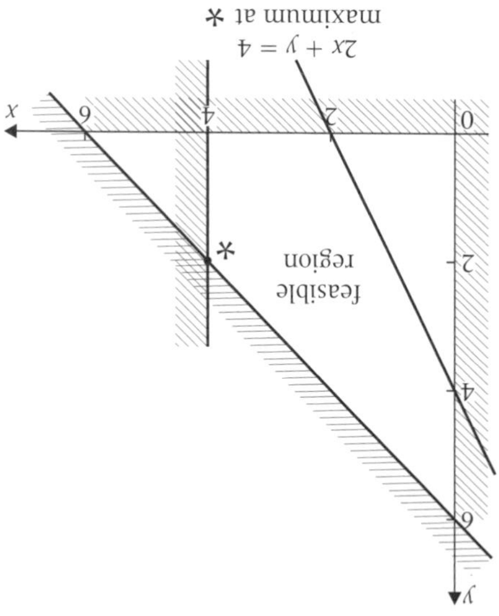

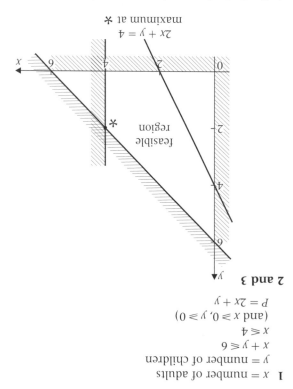

4 $P = 2x + y$ is a maximum at $x = 4$, $y = 2$.
Maximum income $= 2 \times 4 + 2 = £10$ when carrying four adults and two children.

Exam style practice paper

Time allowed: 1 hour 30 minutes
Maximum marks: 80

Answer **all** questions

---

1 A student is using the following algorithm:

LINE 10 Let $A = 1$
LINE 20 Let $B = 1$
LINE 30 Print A, B
LINE 40 Let $C = A + B$
LINE 50 Print C
LINE 60 Let $A = B$
LINE 70 Let $B = C$
LINE 80 If $C < 50$ goto LINE 40
LINE 90 End

(a) Trace the algorithm. *(3 marks)*

(b) Write down the purpose of the algorithm. *(1 mark)*

(c) Another student misread LINE 80. This student used:

LINE 80 If $C < 50$ goto LINE 10 *(2 marks)*

Write down the output that this student would obtain.

(d) Write out an amended algorithm that would print out the first 25 numbers of the series that this algorithm generates. *(3 marks)*

2 A school has five teachers, A, B, C, D and E, available to teach five subjects, maths (M), further maths (F), statistics (S), physics (P) and information technology (I). All five subjects need to be taught at the same time. The subjects each teacher is able to teach are shown in the following table.

Teacher	Subjects
A	I, P
B	I, P
C	I, M, P
D	F, M, S
E	I, M, S

(a) Show this information on a bipartite graph. *(2 marks)*

(b) Initially A is to teach P, B is to teach I, D is to teach M and E is to teach S.

Demonstrate, by using an alternating path from this initial matching, how each teacher can be allocated to a subject that the teacher is able to teach. *(4 marks)*

(c) Teacher D insists on teaching M. Explain why a complete matching is now impossible. *(1 mark)*

3 Use a quick-sort algorithm to rearrange the following numbers into ascending order (you must clearly indicate the pivots that you have used):

23, 2, 31, 4, 30, 11, 5, 13

State the number of comparisons and swaps on each pass of the algorithm. *(8 marks)*

4 Raimondo, an ice-cream salesman, travels along all the roads of a housing estate at least once, starting and finishing at the point A. The lengths, in metres, of the roads are shown in the following diagram.

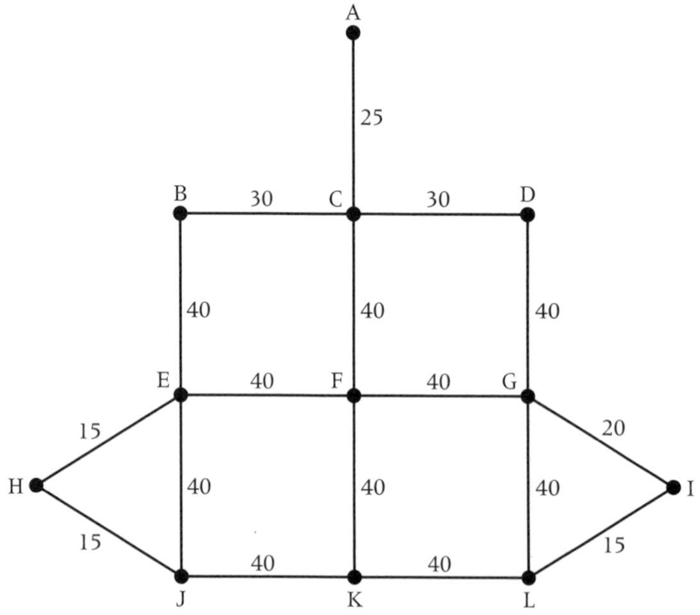

(a) Find the length of an optimal Chinese postman route around the estate. *(6 marks)*

(b) Raimondo drives around another estate that has eight odd vertices. Find the number of ways of pairing these odd vertices. *(2 marks)*

5 The following figure shows a disconnected graph, G, with 10 vertices.

By drawing a new graph in each case, find the minimum number of edges that must be added:

(a) to make graph G connected, *(2 marks)*

(b) to make graph G Eulerian, *(2 marks)*

(c) to make graph G Hamiltonian. *(2 marks)*

6 A large horticultural company is planning a system for watering its plants. The plants are housed in two large greenhouses with water outlets at 14 different points. The following network shows the distances, in metres, between the 14 points that can be connected by water pipes.

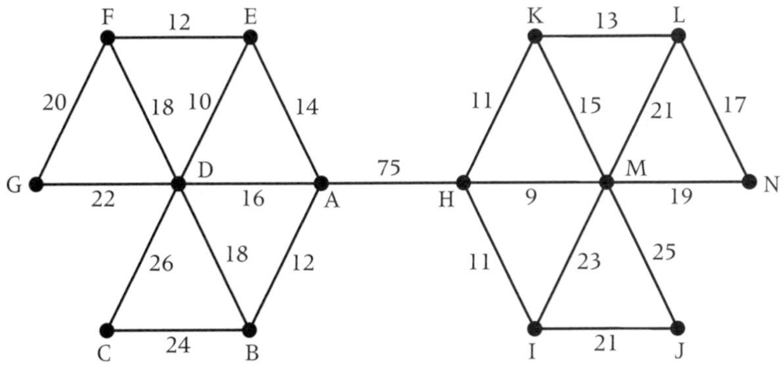

(a) **(i)** Use Kruskal's algorithm to find the minimum length of pipe required to connect the 14 points. *(6 marks)*

Showing your working at each stage:

(ii) draw your minimum spanning tree. *(2 marks)*

(b) The company considers adding a pipe from J to N. The distance JN is x metres. Given that there are now two minimum spanning trees of equal length, find the value of x. *(3 marks)*

7 **(a)** A linear programming problem has the feasible region given by the following inequalities:

$$x \geqslant 2,\ y \geqslant 3,\ x + y \leqslant 20,\ 2x + y \leqslant 24,\ 3x + 2y \geqslant 30$$

Draw a suitable diagram, indicating the feasible region. *(5 marks)*

(b) Use your diagram to find the minimum and maximum values of $x + 3y$ in the feasible region. *(3 marks)*

8 During a general election campaign a politician, who is based at A, has to visit the towns B, C, D and E on a particular day before returning to A. He is trying to find the route that will minimise his travelling distance. The following diagram shows the distances, in kilometres, between the towns.

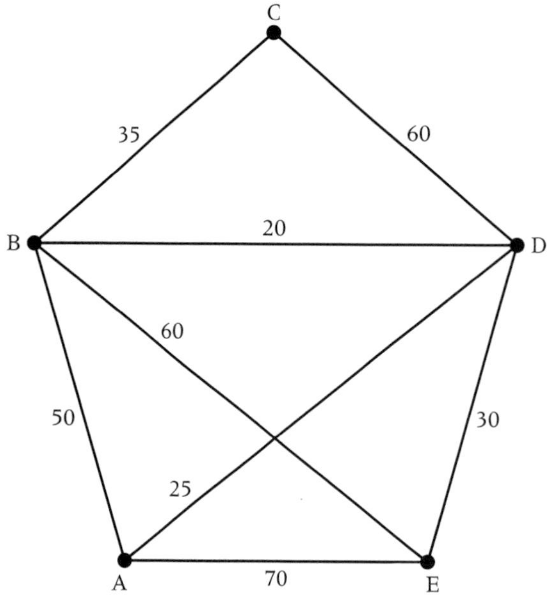

(a) Complete the table below, in which the entries are the **shortest** distances, in kilometres, between pairs of vertices. *(3 marks)*

(b) Use the nearest-neighbour algorithm on this complete network to find an upper bound for the length of a tour of this network that starts and finishes at A. *(4 marks)*

(c) By considering your answer to **(b)**, state the number of times the politician would have to visit each town. *(3 marks)*

	A	**B**	**C**	**D**	**E**
A	–	45		25	
B	45	–	35	20	
C		35	–		
D	25	20		–	30
E				30	–

9 A construction company is to transform a clay mine into a new tropical garden. The company has to bring in three types of soil, X, Y and Z. They find that there must be:

- at least 180 000 tons overall,
- at least 20 000 tons of each type,
- at least 40% of the total must be of type X,
- at most 25% of type Z.

The cost of the soil per ton is £36, £32, £22 for each ton of X, Y and Z, respectively. The company wishes to minimise its expenditure £E, on soil.

The company buys x tons of type X, y tons of type Y and z tons of type Z.
Formulate the above situation as a linear programming problem, simplifying your inequalities so that all the coefficients are integers. *(8 marks)*

Answers

1 Minimum connectors

EXERCISE 1A

1 (a) EG, AC, AB, ED, BE, GF
Total 113

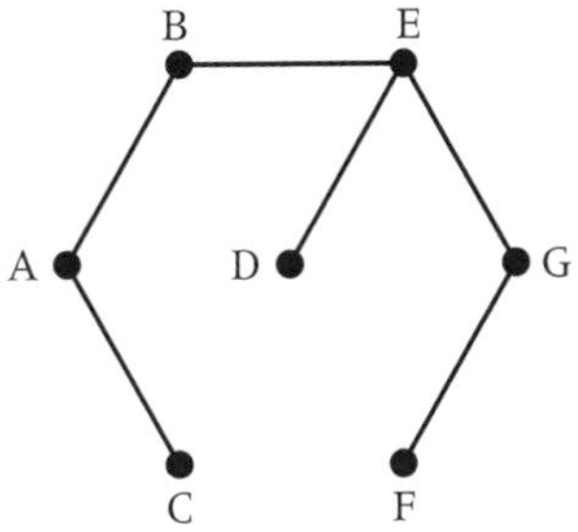

(b) BC, DE, AC, EH, BE, GH, EF, HI
Total 87

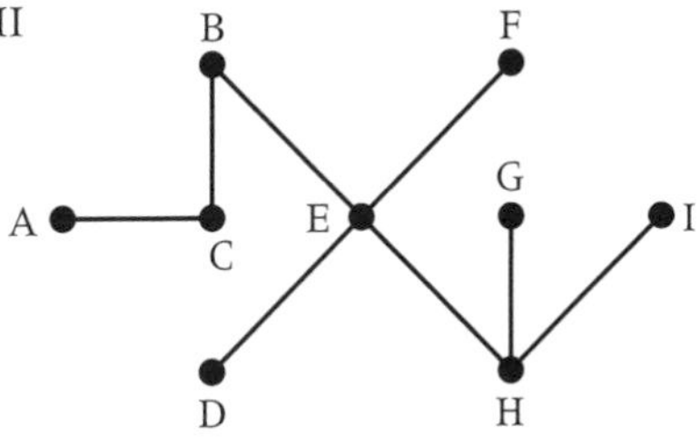

2 RC, GM, RS, GC, GL, DA, DM   Total 33

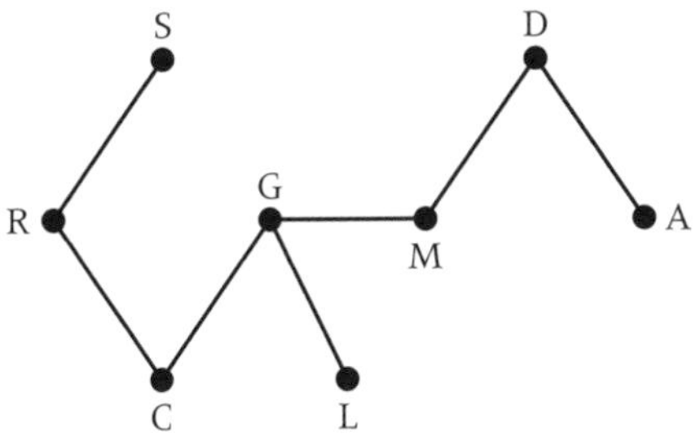

3 FH, JK, EG, BD, KL, AC, IK, AB, GH, DE, HI   Total 257

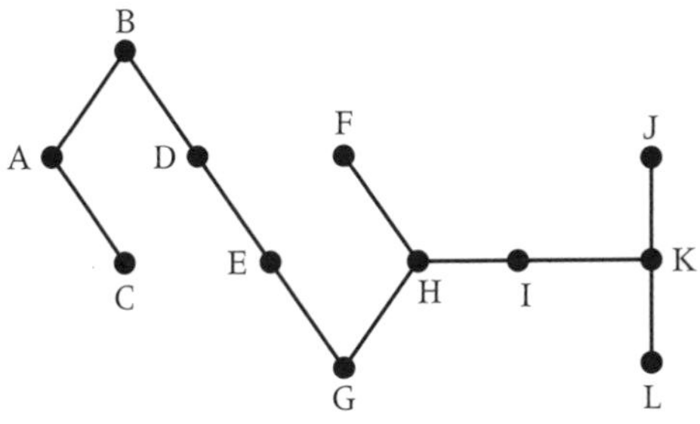

EXERCISE 1B

1 (a) AC, AB, BE, EG, ED, GF
Total 113

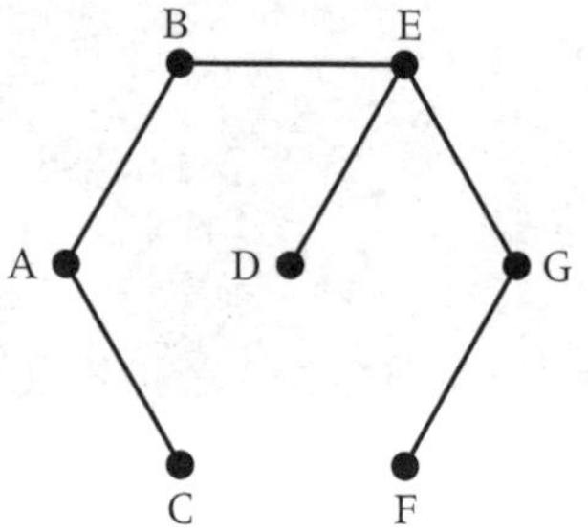

(b) AC, CB, BE, DE, EH, GH, EF, HI
Total 87

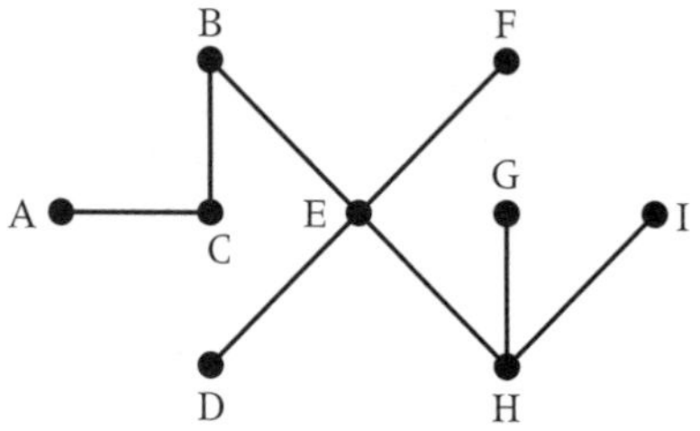

2 AD, DM, MG, GC, RC, RS, GL Total 33

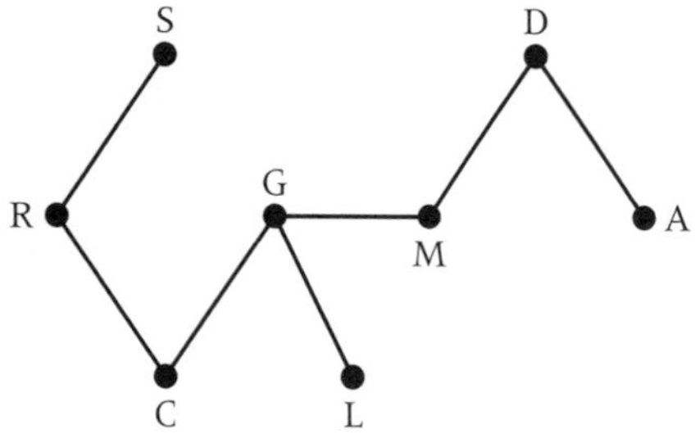

3 AC, AB, BD, DE, EG, GH, FH, HI, IK, KJ, KL Total 257

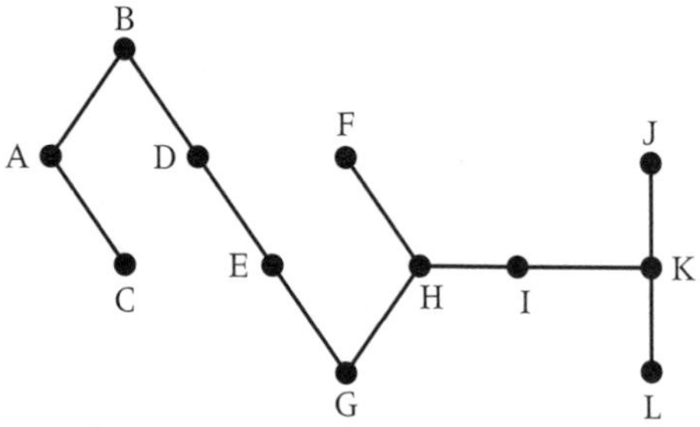

EXERCISE 1C

1 (a) AC, CD, CB, AE
Total 30

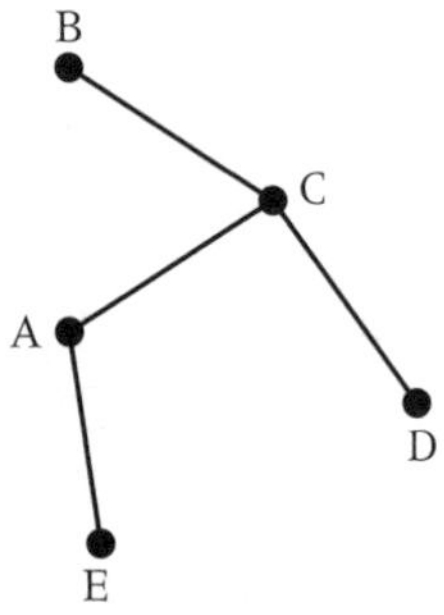

(b) AF, FC, CD, BD, BE
Total 74

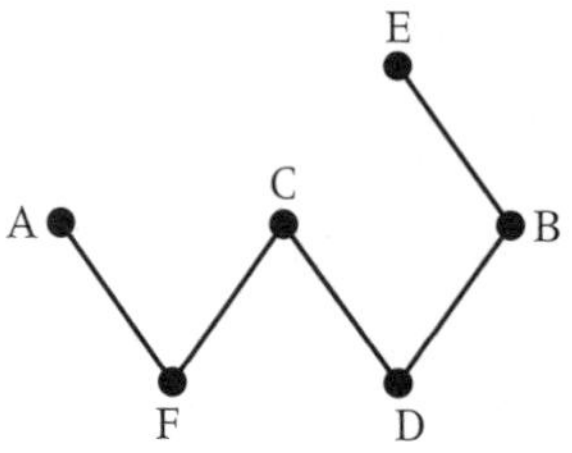

MIXED EXERCISE

1 **(a)** BD, BF, AB, FC, BE **(b)** 52

2 **(a)** $n - 1$

(b) **(i)** AP, PH, AS, SR, RW **(ii)** 15

3 **(a)** AB, BC, BD, CE, EF **(b)** 33

(c) Easier to detect cycles.

4 **(a)** HalHud, BHal, LW, LB, HarL **(b)** 62

5 **(a)** RY, YL, RM, MS, SD, SN, DP; total 171

(b)

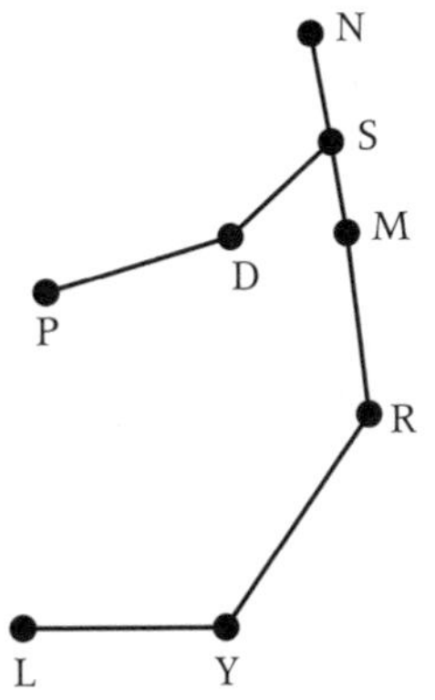

6 BD, FE, HG, BC, AB, GF, DE; total 51.5

7 GF, BD, CD, EG, HJ, AD, IJ, HG, CF; total 36

8 AB, BD, BE, BF, FC; total 525

9 CL, LP, PM, MS, SD, DL, LY, YM and YH; total 271

10 AP, PE, GE, GF, FI; total 336

11 AG, GL, AS, AD, WD; total 365

12 **(a)** AD, AF, FG, AE, EC, BC, EH; total 45

(b)

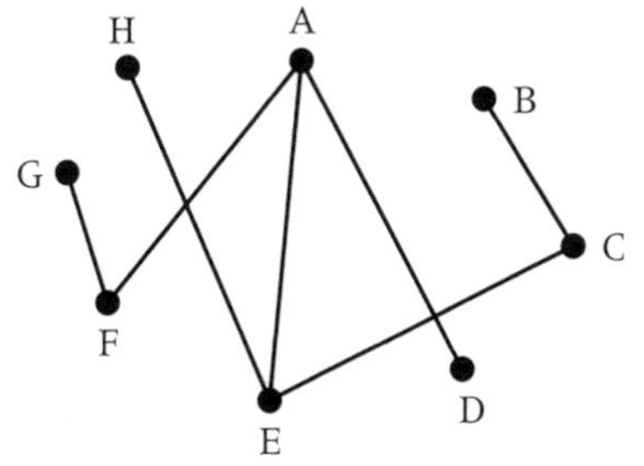

AB = 21

(c) Join AB not BC.

13 EF, EGe, GeDu, GeDa, FS, SP, FI, IGr; total 500

14 **(a)** TG, AH, BC, BH, CD, DG, EG, FG; total 510

(b) TGF, 220

15 (a)

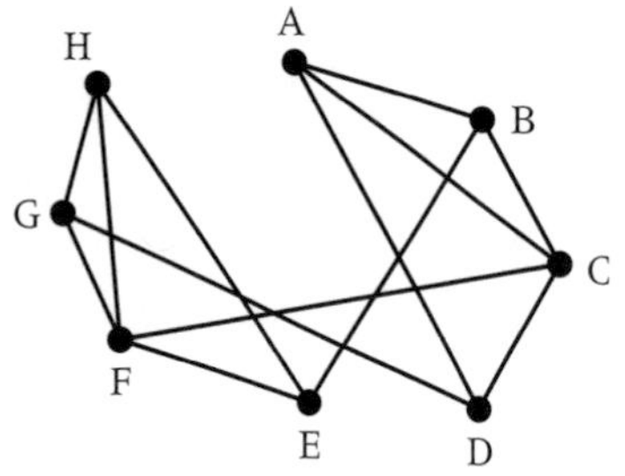

(b) BC, AC, AD, BE, FC, DG, GH; total £69

(c) (i) Row C: 4, 3, 5, 5

(ii) AC, CB, CD, CF, BE, DG, GH; total £57

2 Shortest path problem (Dijkstra's algorithm)

EXERCISE 2A

1 (a)

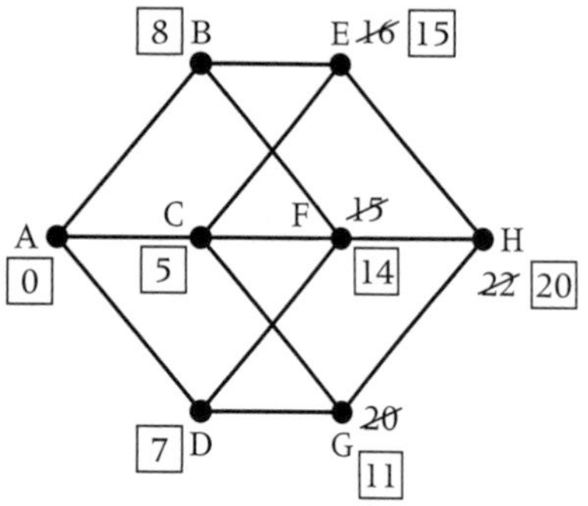

(b)

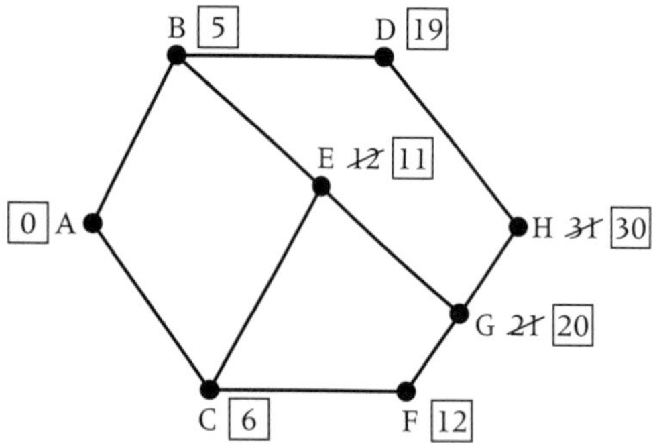

2

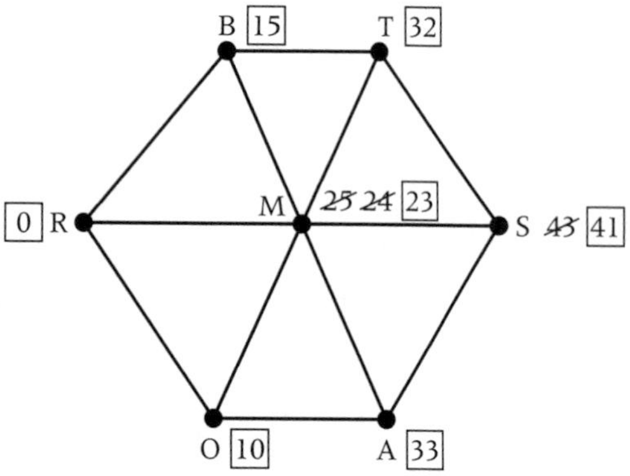

3

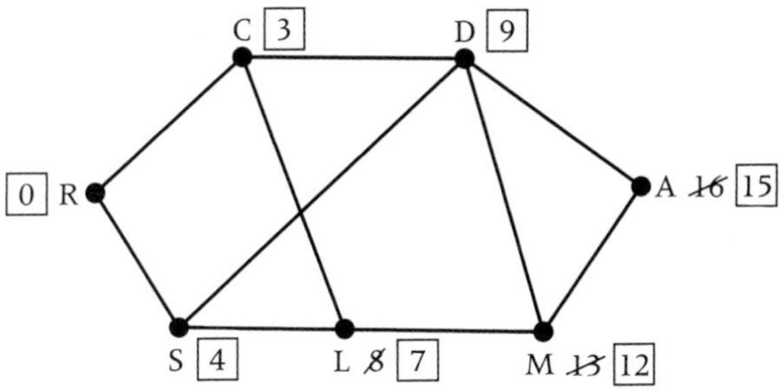

EXERCISE 2B

1 (a)

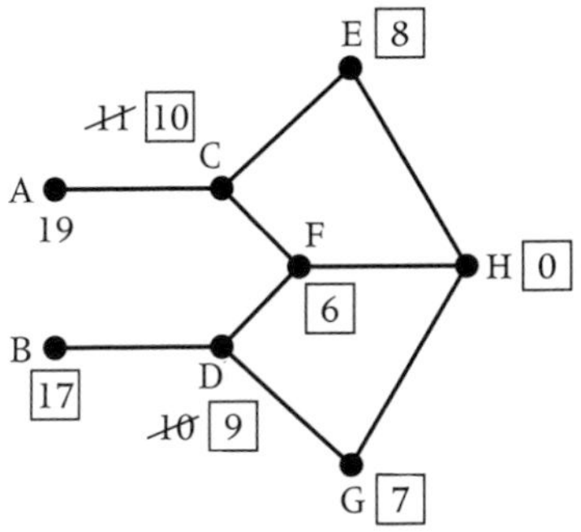

(b)

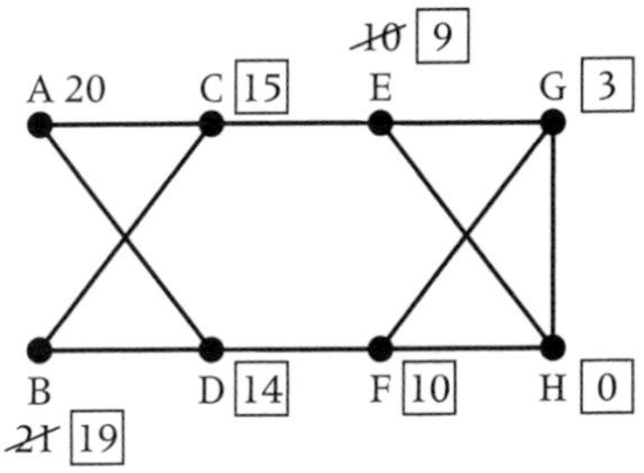

EXERCISE 2C

For 2A

1 (a) ABEH **(b)** ACFGH

2 RBTS

3 RCDMA

For 2B

1 (a) HGDB **(b)** HGECB

MIXED EXERCISE

1 (a) 292, ABEFGHIJ **(b)** DE = 82, ABDEFGHIJ

2 (a) 110 minutes **(b)** 135 minutes

3 26 miles, AEBFGHJKMN

4 (a) 19 miles, ROAST **(b)** 45.75 minutes, RXYT

5 **(a)** 20 minutes **(b)** MPTGB **(c)** 7 minutes

6 J 104 seconds, ADHGKMN
L 100 seconds, BEHGKMN
S 99 seconds, CFEHGKMN

7 ABGHJ, 450 seconds

8 **(a)** 74 minutes, ACGHIL **(b)** $56 + x$

9 **(a)** 130 minutes
(b) **(i)** 133 minutes **(ii)** PCoLA
(c) 43.33 km

10 **(a)** GF, BD, EG, DC, HJ, AD, GF, CF; total 31
(b) **(i)** 26 miles, ADCFGHJ **(ii)** $x = 5$

11 **(a)** AGCBDFE, £56m
(b) **(i)** £17m **(ii)** £45m
(c) £56m, CG, BC, AG, FD, FE, BD
(d) As (c)

12 **(a)** ACEHJ, 37 km
(b) ACGIJ, 42 minutes

13 **(a)** AEHKFGC, 49 minutes
(b) $m = 24$ minutes, AEHDC

14 SWTYZ, 23 km

3 Chinese postman problem

EXERCISE 3A

1 Yes

2 Yes

3 Yes

EXERCISE 3B

1 **(a)** Repeat AFED
Total 33 + 10 = 43

(b) AB + DE = 23
AD + BE = 28
AE + BD = 21
Total 72 + 21 = 93

(c) AE + HI = 31
AH + EI = 38
AI + EH = 28
Total 121 + 28 = 149

EXERCISE 3C

1 **(a)** AFEDABCDEFA **(b)** ABCBGDCDEFAGEFA

(c) ABDCADEHEGHIGFDFGJIGDA

2 ACDFGJIGHIHEGDEDADBA

EXERCISE 3D

1 Repeat DE
Total $93 + 7 = 100$

2 Repeat CE = 12
Total $114 + 12 = 125$

3 Repeat CD + EG
Total $158 + 19 = 177$

MIXED EXERCISE

1 **(a)** **(i)** C, E odd order
(ii) ABCDEBFEDCA, 68 + 12.5 miles = 80.5 miles

(b) **(i)** 15 **(ii)** $(n - 1)(n - 3) \ldots \times 3 \times 1$

2 **(a)** B, D, F, G odd

(b) and **(c)** BD + FG = 5.5, BF + DG = 2.5, BG + DF = 5.5

(d) ABFGDCBFEGEDECA, 22 miles

3 **(a)** **(i)** GS + BW = 19.5, GB + SW = 22, GW + SB = 22.5
(ii) 93 miles

(b) 15 **(c)** $(n - 1)(n - 3) \ldots \times 3 \times 1$

4 **(a)** **(i)** RWCSICRIWCR **(ii)** 1000 m

(b) **(i)** RW **(ii)** 159 m

5 **(a)** and **(b)** AB + CD = 17, AC + BD = 10.5, AD + BC = 11; total 50

(c) 2

6 **(a)** 180 m, ABEBCDEFA

(b) **(i)** B, E
(ii) ABEBCRSRYSDEFQPQXPA, $200 + 10\pi$ m

7 **(a)** Odd order

(b) BE $(2x - 1)$, BAE 13, BCDE $(x + 11)$

(c) 66 minutes, ABCDEBAEA

(d) BCDE

8 **(a)** G1 and G2 – no odd vertices, G3 – yes

(b) 9, 12, 12

9 **(a)** Odd order **(b)** $2.75 < x < 3$

(c) $3x + 210$ minutes, ABFEFDEBCDADECA

(d) $38.5 + x$ minutes, ADEF

10 $x = 9$

11 **(a)** 27 km, ACDEGJ **(b)** 160 km

12 **(a)** AC, AE, EG, GD, BD, GI, IH, CF; total 46 miles

(b) **(i)** ABDGBEGIHFCHECAEIGDBA

(ii) 140 miles

13 **(a)** **(i)** 3 **(ii)** 15

(b) **(i)** 36 **(ii)** AELP **(iii)** 32

14 **(a)** **(i)** A, E, F, G odd order

(ii) 115 m, ADGFEDCABCEBFBA

(b) ABFEGDCA, 50 parking meters

15 395 m, PQRVWSPTUQUVXWTXTP

4 Travelling salesman problem

EXERCISE 4A

1 **(a)** 39, ACBDA
41, BCDAB
39, CBDAC
39, DBCAD
Best 39

(b) 45, ABCDEA
46, BCAEDB
45, CBEDAC
45, DEBCAD
45, EBCADE
Best 45

(c) 50, AFCDBEA
51, BDCEAFB
54, CDBAFEC
54, DCEAFBD
54, ECDBAFE
51, FAECDBF
Best 50

2 **(a)** 34, ACBA
All the same

(b) 28, AEBDCA
29, BDAECB
28, CAEBDC
28, DBEACD
30, EACBDE

(c) 52, AECBDFA
51, BECADFB
54, CAEFBC
54, DCAEFBD
51, ECADFBE
63, FCAEDBBF

EXERCISE 4B

1 **(a)** Del A $7 + 8 + 11 + 12 = 38$
Del B $9 + 11 + 7 + 8 = 35$
Del C $8 + 12 + 7 + 9 = 36$
Best 38

(b) Del A $6 + 7 + 8 + 8 + 9 = 38$
Del B $13 + 8 + 9 + 6 + 7 = 43$
Del C $7 + 8 + 8 + 6 + 13 = 42$
Del D $6 + 7 + 8 + 8 + 10 = 39$
Del E $6 + 8 + 10 + 7 + 8 = 39$
Best 43

(c) Del A $7 + 8 + 8 + 9 + 7 + 8 = 47$
All 47

2 **(a)** All 34

(b) Del A $4 + 6 + 7 + 3 + 5 = 25$
All 25

(c) Del A $8 + 7 + 8 + 9 + 8 + 9 = 49$
All 49

3 **(a)** $38 \leqslant T \leqslant 39$

(b) $43 \leqslant T \leqslant 45$

(c) $47 \leqslant T \leqslant 50$

EXERCISE 4C

1 Upper bounds:
ABCGDEFA 66
BCGADEFB 71
CGABDEFC 57
DEFCGABD 57
EFDCGABE 63
FEDCGABF 61

Lower bounds:
Del A 47
Del B 47
Del C 52
Del D 46
Del E 43
Del F 46

$52 \leqslant T \leqslant 57$
Tour of 57 is CGABCDEFC

2 Upper bounds:
AICBHGFEDA 102
BCAIHGFEDB 102
CBAIHGFEDC 94
DEFGHCBAID 94
EFGHCBAIDE 94
FEDCBAIHGF 94
GFEDCBAIHG 94
HCBAIDEFGH 94
IABCHGFEDI 94

Lower bounds:
Del A 81
Del B 79
Del C 86
Del D 83
Del E 83
Del F 83
Del G 83
Del H 83
Del I 81

$86 \leqslant T \leqslant 94$
Tour of 94 is CBAICHGFEDC

3 Upper bounds:
ACFEDBGA 76
BCAFEDGB 80
CABFEDGC 75
DCABFEGD 88
EDCABFGE 84
FCABDEGF 84
GCABFEDG 75

Lower bounds:
Del A
Del B 83
Del C 9
Del D 8
Del E 94
Del F 88
Del G 85

MIXED EXERCISE

1 **(a)** RC, RW, RL, RMn, CS, RMi; total 31 minutes

(b) RCSWLMiMnR; total 57 minutes

(c) 44 minutes

2 **(a)** RLDCOBR, 625

(b) 505

3 **(a)** AT, TN, TC, ST, NB, BH; 81

(b) **(i)** 95

(ii) AT and AC don't make a cycle

4 Del M – 355 km
Del L – 355 km
Del S – 362 km
Del B – 368 km

5 **(a)** **(i)** 338 seconds

(ii) Exists, may be improved

(b) O13542O, 330 seconds

(c) 305 seconds

6 **(a)** AD, AB, BE, EF, EC; 72 miles

(b) Del A – 104 miles
Del B – 88 miles
Del C – 88 miles
Del D – 104 miles

7 **(a)** MSCTPBM, 116

(b) 112

(c) $112 \leqslant T \leqslant 116$

8 **(a)** ADCBA, 320 km

(b) Exists, may be improved

9 **(a)** EY, BC, CD, DE, DZ, AX, AB, 47 miles

(b) 94 miles

(c) **(i)** ZDCBAXEYZ, 64 miles
(ii) YEDCBAXEYZY, 78 miles

(d) 61 miles

(e) $61 \leqslant L \leqslant 64$

10 **(a)** **(i)** 131 minutes
(ii) Exists, may be improved

(b) ABDGFECA, 127 minutes

(c) 114 minutes

11 **(a)** **(i)** AFBCDEA, 165 minutes
(ii) Exists, may be improved

(b) **(i)**

–	25	25	0	15	23
20	–	30	20	0	5
0	15	–	5	20	15
10	45	25	–	15	0
15	55	15	5	–	0
0	20	45	15	10	–

–	10	10	0	15	23
20	–	15	20	0	5
0	0	–	5	20	15
10	30	10	–	15	0
15	40	0	5	–	0
0	5	30	15	10	–

135 minutes

(ii) CADFBEC

12 **(a)** **(i)** SPOBLRS, 85 minutes
(ii) ROBPSLR, 82 minutes

(b) Exists, may be improved

(c) SROBLPS, 78 minutes

13 **(a)** LA, SB, LV, PS, SD, LA; 745 miles

(b) 745 miles

(c) $T = 745$ miles

14 **(a)** DFGEJIHD, 53 minutes

(b) FG, GE, EJ, JI, IH, 40 minutes

(c) Lower bound also equals 53 minutes

15 **(a)** AEBDCA, 42 miles

(b) EB, BD, DC; 26 miles

(c) $36 \leqslant T \leqslant 42$

5 Graph theory

MIXED EXERCISE

1 **(a)** Cycle visiting every vertex once before returning to the start vertex.

(b) 3

(c) $(n - 1)!/2$

2 **(a)**

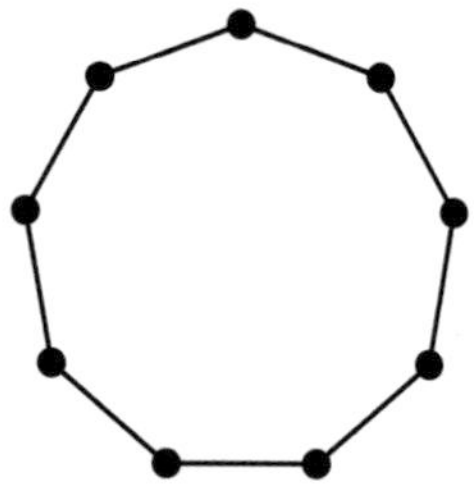

(b) **(i)** 2, 4, 6, 8

(ii) 9, 18, 27, 36

3 **(a)** Needs 10 edges

(b) **(i)** 30

(ii) May be loops or cycles

(iii)

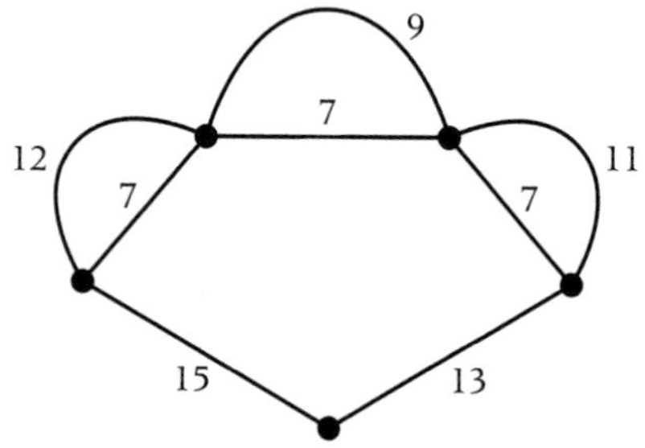

4 (a)

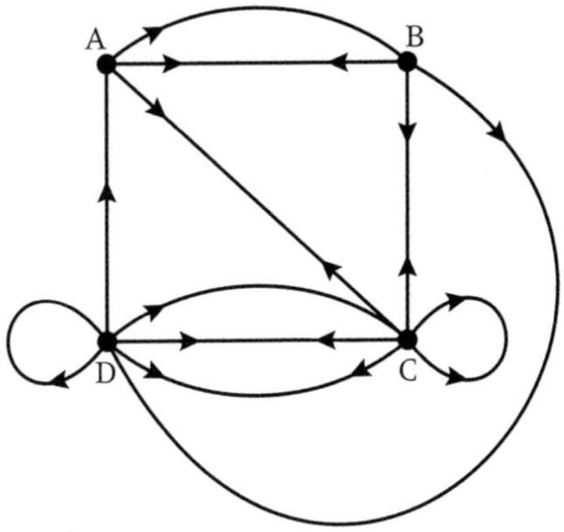

(b) Entries not symmetrical

5 (a) (i)

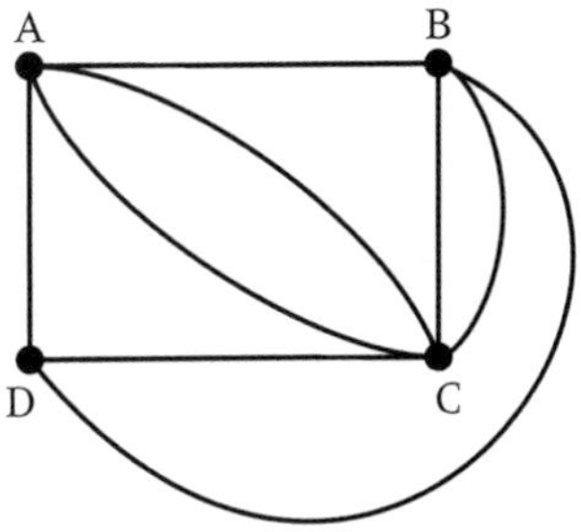

(ii) C, D odd

(b) Sum = 2 × number of edges

(c) Directed graph

6 (a) 3

(b) (i) $n - 1$ **(ii)** $(n - 1)!/2$

(c)

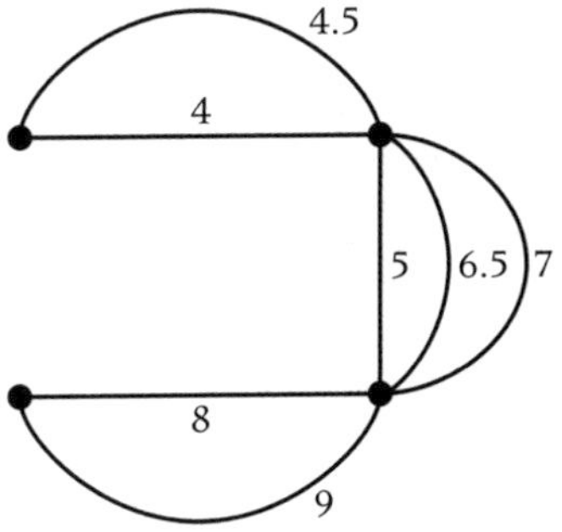

7 (a) 25 **(b)** 33 **(c)** 62

(d)

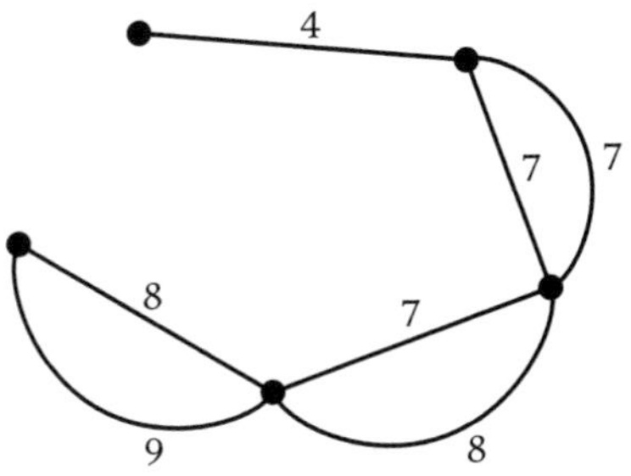

8 (a) $5 + (49 \times 7) + 14 + (7 \times 2) = 376$

(b) 34

(c) $mn + m + 3$

(d) $61/60x + 3$

9 (a) 2, 4, 6; edges 7, 14, 21

(b)

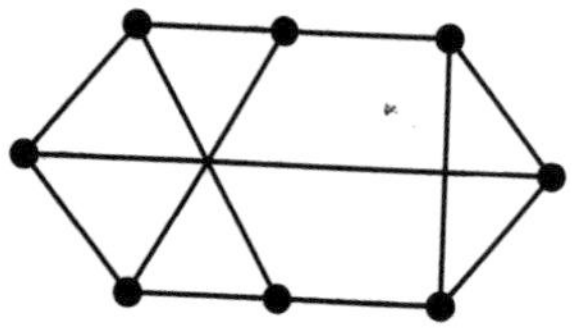

$d = 2, 3, 4, 5, 6, 7$

10 (a) 24

(b) 49!, 1.93×10^{48}

11 (a) (i) 3

(ii) 6

(b) (i) $n - 1$

(ii) $n(n - 1)/2$

12 (a)

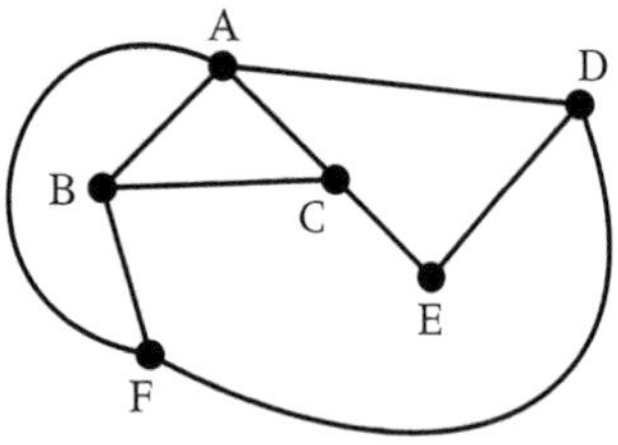

(b) Neither, four odd vertices

(c) AFDECBA

(d) Hamiltonian. Want to visit all places as quickly as possible.

13 (a) (i) Y, Z are odd

(ii) ZYTWUZWYVSTVXY

(b) ZUWTSVXYZ

(c) (i) YZ

(ii) ZUW forms a cycle

14

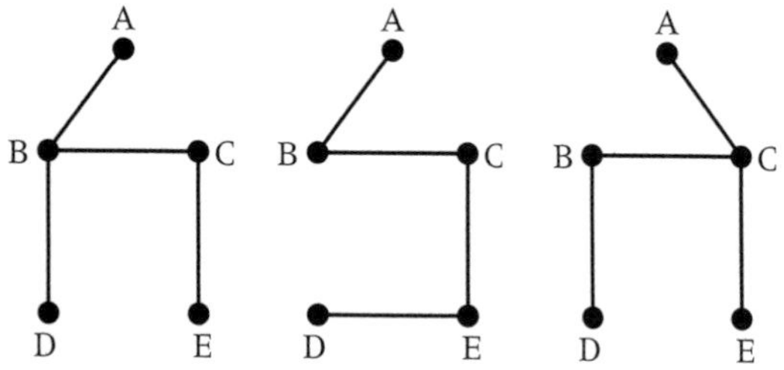

6 Matchings

EXERCISE 6A

1

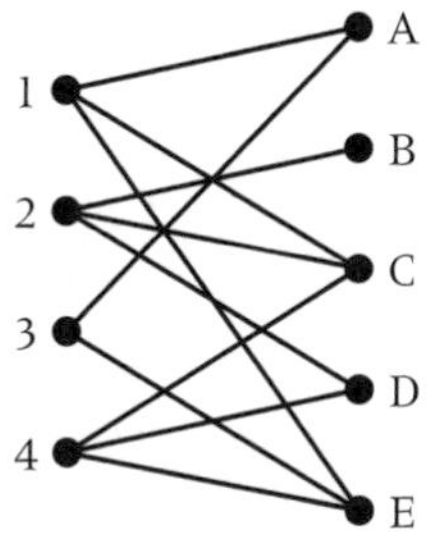

2

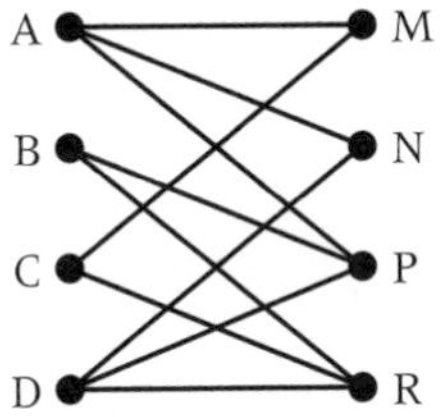

3

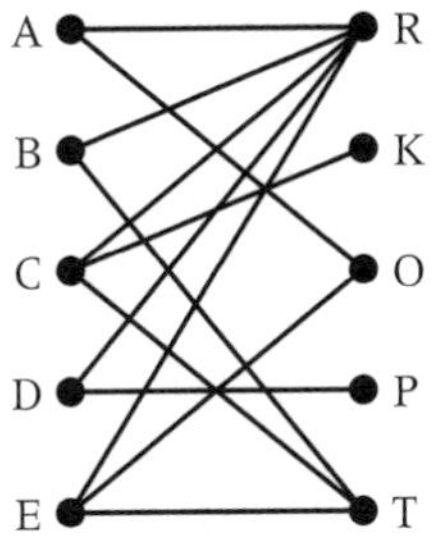

4

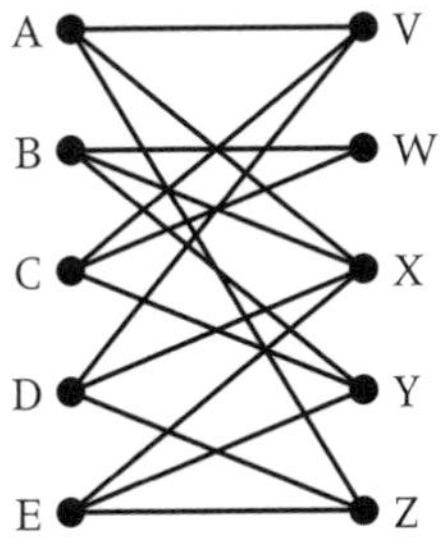

EXERCISE 6B

1 D and C can only be matched to 4.

2 (A1, B2, C3, D4), (A2, B1, C3, D4), (A1, B2, C4, D3), (A2, B1, C4, D3)

3 (A1, B2, C3, D4)

EXERCISE 6C

1 $B - 2 \not- D - 3$ (A1, B2, C4, D3)

2 $E - R \not- D - S$ (AQ, BT, CP, DS, ER)

3 Both D and E are missing from initial match.
$D - 5$, $E - 6 \not- F - 1$ (A2, B3, C4, D5, E6, F1)

MIXED EXERCISE

1 (a)

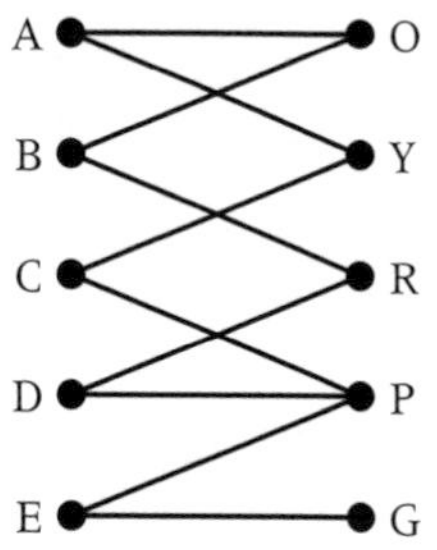

(b) B – O ≠ A – Y ≠ C – P ≠ E – G
AY, BO, CP, DR, EG

2 (a)

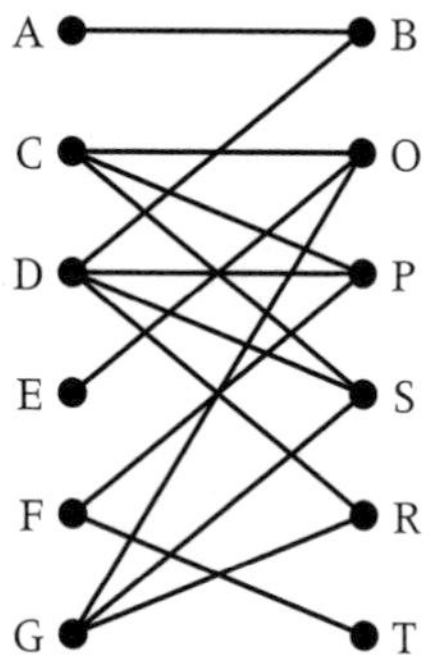

(b) A – B ≠ D – R, E – O ≠ C – P ≠ F – T
AB, CP, DR, EO, FT, GS

3 (a)

	A	B	C	D
1	1	1	0	0
2	1	1	1	0
3	0	0	1	1
4	0	0	1	1

(b) (i) 3 ≠ D – 4 ≠ C – 2 ≠ B; 1A, 2B, 3D, 4C

(ii) 3 – C ≠ 2 – B; 1A, 2B, 3C, 4D
3 – D ≠ 4 – C ≠ 2 – A ≠ 1 – B; 2A; 1B, 4C, 3D

4 (a)

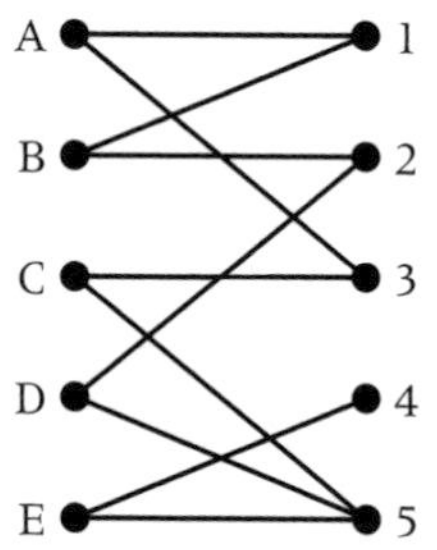

(b) B – 1 ≠ A – 3 ≠ C – 5 ≠ E – 4; A3, B1, C5, D2, E4

5 (a)

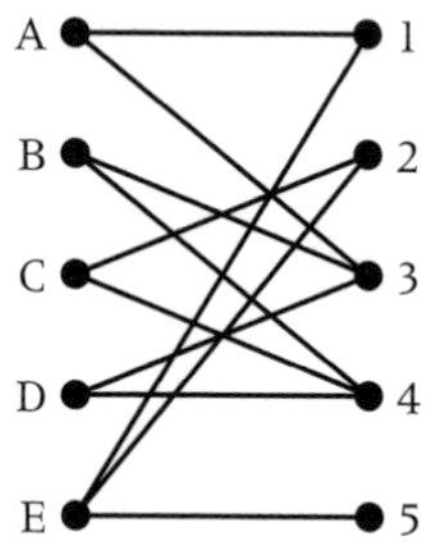

(b) C – 2, D – 3 ≠ A – 1 ≠ E – 5;
A1, B4, C2, D3, E5

6 (a)

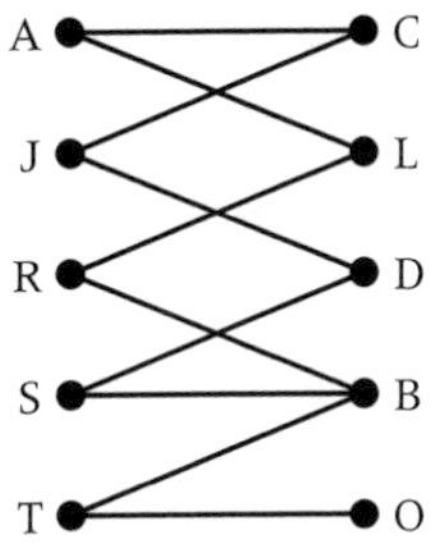

(b) J – C ≠ A – L ≠ R – B ≠ T – O; AL, JC, RB, SD, TO

7 (a)

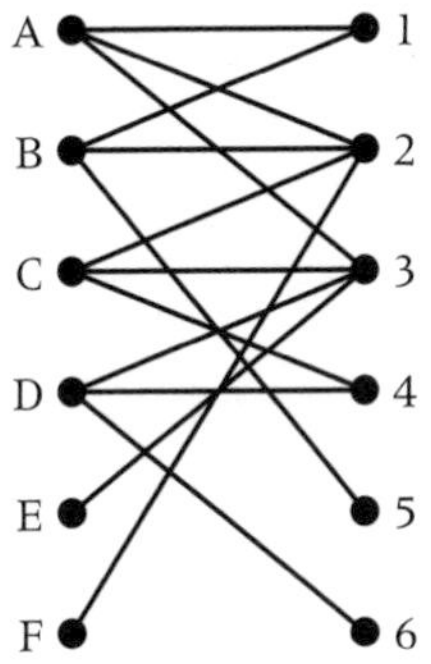

(b) E – 3 ≠ C – 2 ≠ A – 1 ≠ B – 5, F – 2 ≠ C – 4 ≠ D – 6; A1, B5, C4, D6, E3, F2

7 Sorting algorithms

EXERCISE 7A

1 (a)

18	3	45	17	1	26	43	22	16
3	18	17	1	26	43	22	16	45
3	17	1	18	26	22	16	43	45
3	1	17	18	22	16	26	43	45
1	3	17	18	16	22	26	43	45
1	3	17	16	18	22	26	43	45
1	3	16	17	18	22	26	43	45
1	3	16	17	18	22	26	43	45

(b)

18	3	45	17	1	26	43	22	16
18	45	17	3	26	43	22	16	1
45	18	17	26	43	22	16	3	1
45	18	26	43	22	17	6	3	1
45	26	43	22	18	17	16	3	1
45	43	26	22	18	17	16	3	1

2

32	11	3	27	16	9	23	19	C	S
11	3	27	16	9	23	19	32	7	7
3	11	16	9	23	19	27	32	6	5
3	11	9	16	19	23	27	32	5	2
3	9	11	16	19	23	27	32	4	1
3	9	11	16	19	23	27	32	3	0

3

R	L	C	A	B	D	W	H	C	S
L	C	A	B	D	R	H	W	7	6
C	A	B	D	L	H	R	W	6	5
A	B	C	D	H	L	R	W	5	3
A	B	C	D	H	L	R	W	4	0

EXERCISE 7B

1 (a)

18	3	45	17	1	26	43	22	16
3	18	45	17	1	26	43	22	16
3	18	45	17	1	26	43	22	16
3	17	18	45	1	26	43	22	16
1	3	17	18	45	26	43	22	16
1	3	17	18	26	45	43	22	16
1	3	17	18	26	43	45	22	16
1	3	17	18	22	26	43	45	16
1	3	16	17	18	22	26	43	45

(b)

18	3	45	17	1	26	43	22	16
18	3	45	17	1	26	43	22	16
45	18	3	17	1	26	43	22	16
45	18	17	3	1	26	43	22	16
45	18	17	3	1	26	43	22	16
45	26	18	17	3	1	43	22	16
45	43	26	18	17	3	1	22	16
45	43	26	22	18	17	3	1	16
45	43	26	22	18	17	16	3	1

2

32	11	3	27	16	9	23	19	C	S
11	32	3	27	16	9	23	19	1	1
3	11	32	27	16	9	23	19	2	2
3	11	27	32	16	9	23	19	2	1
3	11	16	27	32	9	23	19	3	2
3	9	11	16	27	32	23	19	5	4
3	9	11	16	23	27	32	19	3	2
3	9	11	16	19	23	27	32	4	3

3

R	L	C	A	B	D	W	H	C	S
L	R	C	A	B	D	W	H	1	1
C	L	R	A	B	D	W	H	2	2
A	C	L	R	B	D	W	H	3	3
A	B	C	L	R	D	W	H	4	3
A	B	C	D	L	R	W	H	3	2
A	B	C	D	L	R	W	H	1	0
A	B	C	D	H	L	R	W	3	2

EXERCISE 7C

1 (a)

18	3	45	17	1	26	43	22	16
1	3	43	17	16	26	45	22	18
1	3	16	17	18	22	43	26	45
1	3	16	17	18	22	26	43	45

(b)

18	3	45	17	1	26	43	22	16
18	26	45	22	16	3	43	17	1
45	26	43	22	18	17	16	3	1
45	43	26	22	18	17	16	3	1

2

32	11	3	27	16	9	23	19	C	S
16	9	3	19	32	11	23	27	4	3
3	9	16	11	23	19	32	27	8	3
3	9	11	16	19	23	27	32	10	3

3

R	L	C	A	B	D	W	H	C	S
B	D	C	A	R	L	W	H	4	2
B	A	C	D	R	H	W	L	7	2
A	B	C	D	H	L	R	W	10	4

EXERCISE 7D

1 (a)

18	3	45	17	1	26	43	22	16
3	17	1	16	18	45	26	43	22
1	3	17	16	18	26	43	22	45
1	3	16	17	18	22	26	43	45

(b)

18	3	45	17	1	26	43	22	16
45	26	43	22	18	3	17	1	16
45	26	43	22	18	17	16	3	1
45	43	26	22	18	17	16	3	1

2

32	11	3	27	16	9	23	19	C	S
11	3	27	16	9	23	19	32	7	7
3	9	11	27	16	23	19	32	6	2
3	9	11	16	23	19	27	32	4	3
3	9	11	16	23	19	27	32	2	0
3	9	11	16	19	23	27	32	1	6

3

R	L	C	A	B	D	W	H	C	S
L	C	A	B	D	H	R	W	7	6
C	A	B	D	H	L	R	W	5	5
A	B	C	D	H	L	R	W	4	2
A	B	C	D	H	L	R	W	2	0

MIXED EXERCISE

1 (a)

4	7	13	26	8	15	6	56
4	7	13	8	15	6	26	56
4	7	8	13	6	15	26	56
4	7	8	6	13	15	26	56
4	7	6	8	13	15	26	56
4	6	7	8	13	15	26	56

(b) 28

2

R	I	W	D	P	L	A
I	D	P	L	A	R	W
D	A	I	P	L	R	W
A	D	I	L	P	R	W

3

14	27	23	36	18	25	16	66
14				18			
	25				27		
		16				23	
			36				66
14	25	16	36	18	27	23	66
14		16		18		23	
	25		27		36		66
14	25	16	27	18	36	23	66
14	16	18	23	25	27	36	66

4

N	E	D	P	A
E	D	N	A	P
D	E	A	N	P
D	A	E	N	P
A	D	E	N	P

5 (a)

9	5	7	11	2	8	6	17
5	7	2	8	6	9	11	17
2	5	7	8	6	9	11	17
2	5	6	7	8	9	11	17

(b) (i) 28

(ii) $n(n-1)/2$

6

63	32	70	26	59	41	17
32	26	59	41	17	63	70
26	17	32	59	41	63	70
17	26	32	41	59	63	70

7 (a)

P	B	R	G	O	W
B	G	O	P	R	W
B	G	O	P	R	W
B	G	O	P	R	W
B	G	O	P	R	W

(b) 15

(c) $n(n-1)/2$

8 Algorithms

EXERCISE 8B

1

A	*B*	Print
1	1	1
2	8	8
3	27	27
4	64	64
5	125	125
6	216	216
7	343	343
8	512	512
9	729	729
10	1000	1000
11	1331	

2 (a)

A	*B*	*C*	*D*	Print
20	24	1	4	4
4	20	5	0	

(b)

A	*B*	*C*	*D*	Print
72	60	1	12	12
60	12	5	0	

3

A	*B*	*C*	Print
27	15	0	
26		15	
13	30		
12		45	
6	60		
3	120		
2		165	
1	240		
0		405	405

EXERCISE 8C

1
```
LINE 10  For I = 1 to 10
LINE 20  Let J = I*2
LINE 30  Print I, J
LINE 40  Next I
LINE 50  End
```

2
```
LINE 10  Let A = 1
LINE 20  Let B = 0
LINE 30  If B > 99 then goto LINE 80
LINE 40  Let B = A*A
LINE 50  Print B
LINE 60  Let A = A + 1
LINE 70  Goto LINE 30
LINE 80  End
```

3
```
LINE 10  For M = 10 to 50
LINE 20  For N = 2 to M − 1
LINE 30  If M/N = Int(M/N) then goto LINE 60
LINE 40  Next N
LINE 50  Print 'M is prime'
LINE 60  Next M
LINE 70  End
```

4
```
LINE 10  For I = 1 to 10
LINE 20  For J = 1 to 10
LINE 30  Let K = I*J
LINE 40  Print I*J = K
LINE 50  Next J
LINE 60  Next I
LINE 70  End
```

EXERCISE 8D

1 (a)
```
LINE 10  For I = 5 to 10
LINE 20  Let J = 3*I
LINE 30  Print I, J
LINE 40  Next I
LINE 50  End
```

(b)
```
LINE 10  Let A = 5
LINE 20  Let B = 3*A
LINE 30  If A > 10 then goto LINE 70
LINE 40  Print A, B
LINE 50  Let A = A + 1
LINE 60  Goto LINE 20
LINE 70  End
```

2 (a)
```
LINE 10  For N = 50 to 100
LINE 20  If sqrt(N) ≠ Int(sqrt(N)) then goto LINE 40
LINE 30  Print sqrt(N)
LINE 40  Next N
LINE 50  End
```

(b)
```
LINE 10  Let A = 50
LINE 20  Let B = sqrt(A)
LINE 30  If B ≠ Int(sqrt(A)) then goto LINE 50
LINE 40  Print B
LINE 50  Let A = A + 1
LINE 60  If A > 100 then goto LINE 70
LINE 70  Goto LINE 30
LINE 80  End
```

3 **(a)**

```
LINE 10   For I = 1 to 10
LINE 20   For J = 1 to 10
LINE 30   Let K = I*J
LINE 40   Next J
LINE 50   Next I
LINE 60   End
```

(b)

```
LINE 10    Let A = 1
LINE 20    LET B = 1
LINE 30    Let C = A*B
LINE 40    Print A*B = C
LINE 50    Let A = A + 1
LINE 60    If A < 11 then goto LINE 30
LINE 70    Let A = 1
LINE 80    Let B = B + 1
LINE 90    Goto LINE 30
LINE 100   End
```

MIXED EXERCISE

1 **(a)**

A	*B*	Print
1	1	1, 1
2	8	2, 8
3	27	3, 27
4	64	4, 64
5	125	

(b) *A* would swap from 1 to 2 and *B* would always be 1. The algorithm would not stop.

2 **(a)** **(i)** Maximum mark **(ii)** Next *I*
(iii) Stop

(b) Delete LINE 2; LINE 3.5 If Mark = −1 then goto LINE 7
LINE 6 goto LINE 3

3

X	*Y*
24	20
4	16
12	4
8	
4	

4 **(a)** **(i)** Zeroing **(ii)** Nested loop
(iii) To count the number of calculations.

(b) $I = 2, J = 3, K = 6$

5 **(a)**

Q	*R*	*Store*
7	9	9
3	18	36
1	36	63

(b)

Q	*R*	*Store*
4	8	
2	16	
1	32	32

(c) 10 reduces to 1 faster than 25 reduces to 1.

6 **(a)**

41	23	12	45	17	11	26	58	3	24
23	12	17	11	26	3	24	41	45	58
12	17	11	3	23	26	24	41	45	58
11	3	12	17	23	24	26	41	45	58
3	11	12	17	23	24	26	41	45	58

(b) 45

(c) 58 45 41 26 24 23 17 12 11 3

7 **(a)** **(i)**

A	B	C	D	X_1	X_2
1	−4	4	0		
				2	
					2

(ii)

A	B	C	D	X_1	X_2
2	9	9	9		
				−1.5	
					−3

(b) **(i)** $A = 1, B = 1, C = 1$

(ii) LINE 15 If $A = 0$ then goto LINE 110
LINE 20 If $D < 0$ then goto LINE 110

8 **(a)**

A	I
0	1
200	
216	2
416	
449.28	3
649.28	
701.22	

(b) Swap Next I, Print A

(c)
```
Input P
For I = 1 to N
   A = A + P
   A = (1 + R/100) × A
   Next I
Print A
Stop
```

9 **(a)** **(i)**

X	Y	A	B
5	20	20	0
		15	1
		10	2
		5	3
		0	4

(ii)

X	Y	A	B
7	29	29	0
		22	1
		15	2
		8	3
		1	4

(b) Divides Y by X to give quotient B and remainder A.

10 (a)

A	B	N	S	H	X	Y
1	3	2	0	1	1	
					1.5	2.25
			2.25			
					2	
		1			2.5	6.25
			8.5			
					3	
		0				

(b)

A	B	N	H	X	P	Q	R	Y	S
1	3	4	0.5	1	1				
				1.5		2.25			
				2			4	14	7/3
		2			4				
				2.5		6.25			
				3			9	38	26/3
		0							

9 Linear programming

EXERCISE 9A

1

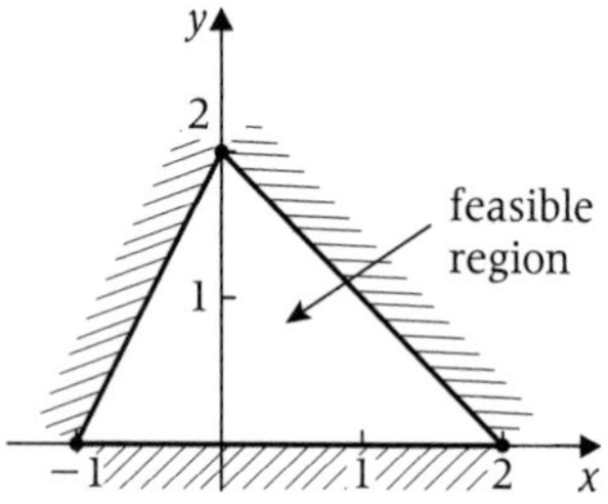

2

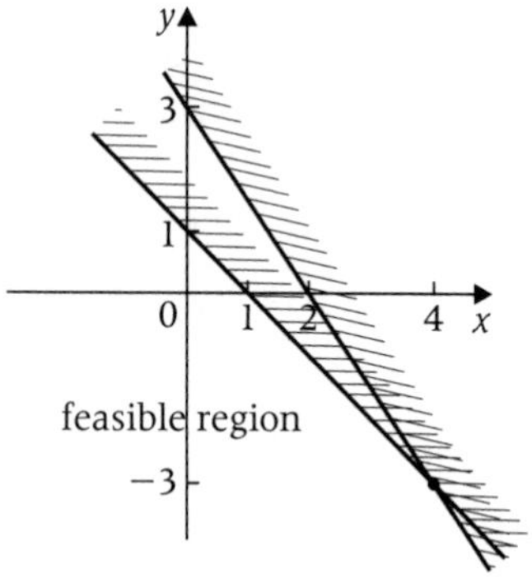

3

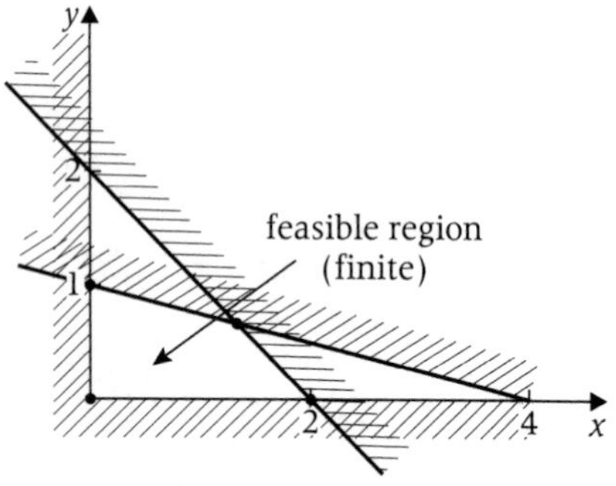

EXERCISE 9B

1 If x = no. of crates of beer and y = no. of crates of wine:

maximise $P = 10x + 15y$
subject to $x + y \leq 4$
$10x + 20y \leq 60$
$x \geq 0$ and $y \geq 0$

2 **(a)** If x = no. of blouses and y = no. of skirts:

maximise $P = 8x + 6y$
subject to $x + y \leq 7$
$x + \frac{1}{2}y \leq 5$
$x \geq 0$ and $y \geq 0$

(b) For example, $x = 2$, $y = 4$; $x = 3$, $y = 4$, etc.

3 If x = no. of large vans and y = no. of small vans:

minimise $C = 40x + 20y$
subject to $5x + 2y \geq 30$
$2x + y \leq 15$
$x \leq y$
$x \geq 0$ and $y \geq 0$

4 If x = no. of boxes of woodscrews and y = no. of boxes of metal screws:

maximise $P = 10x + 17y$
subject to $3x + 2y \leq 3600$
$2x + 8y \leq 3600$
$x \geq 0$ and $y \geq 0$

5 If x = no. of unskilled workers and y = no. of skilled workers:

$x + 2y \leq 180$
$x + y \geq 110$
$y \geq 40$
$y \geq \frac{1}{2}x$
$x \geq 0$

EXERCISE 9C

1 $x = 2$, $y = 2$ for max. $P = 50$.

2 $x = 3$, $y = 4$, for max. $P = 48$.

3 $x = 4$, $y = 5$ for min. $C = 260$.

4 $x = 1080$, $y = 180$ for max. $P = 13\,860$.

5 **(a)** 70 (with 40 unskilled workers).

(b) 90 (with 45 skilled workers).

6 **(a)**

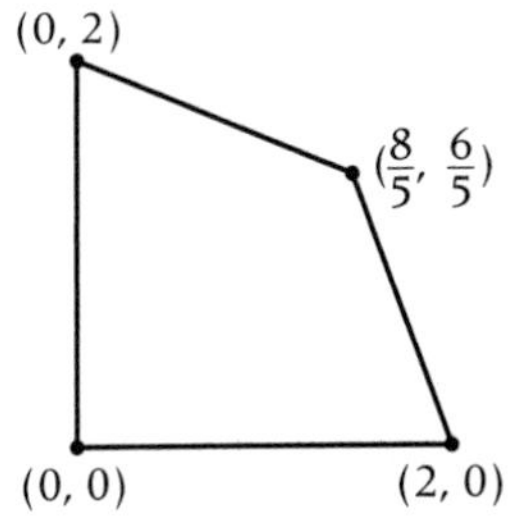

(b) P at (2, 0), Q at $(\frac{8}{5}, \frac{6}{5})$, R at (0, 2).

7 Type A = 819, type B = 162 with profits of £376.20 (best integer solution).

MIXED EXERCISE

1 (a) $800 \leqslant x + y \leqslant 900$, $2x + 3y \leqslant 2200$, $x \geqslant 300$, $y \geqslant 300$, $T = 20x + 25y$

(b)

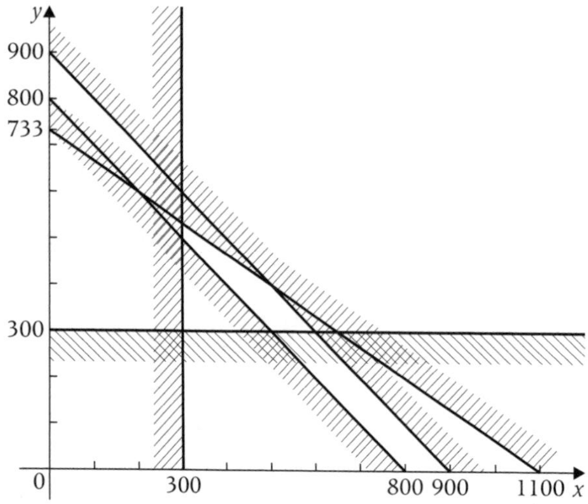

(c) Min. £17 500; max. £20 000

2 (a) $2 \leqslant x \leqslant 20$, $2 \leqslant y \leqslant 20$, $8 \leqslant x + y \leqslant 24$

(b) (i) 64

(ii) −16

3 (a) $x \geqslant 2$, $y \geqslant 5$, $z \geqslant 2$, $x + y + z \geqslant 28$, $40 \leqslant 3x + 2y + 2z \leqslant 60$

(b) (i) Put $x = z$

(ii)

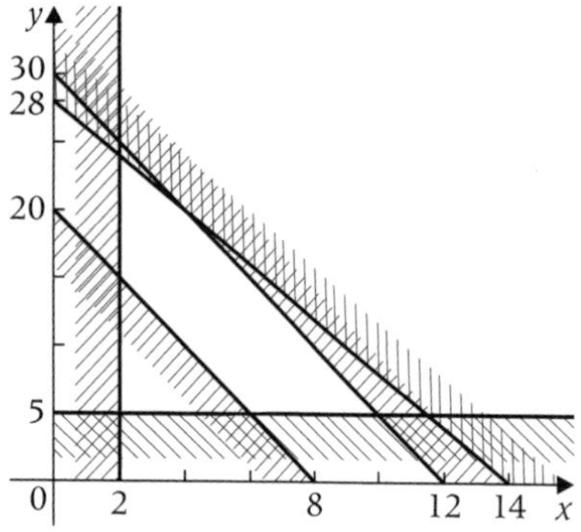

(iii) (2, 24, 2), (3, 22, 3), (4, 20, 4), total 28

4 (a) $x \geqslant 1, y \geqslant 1, 6 \leqslant 3x + y \leqslant 12, 2x + y \geqslant 5$

(b)

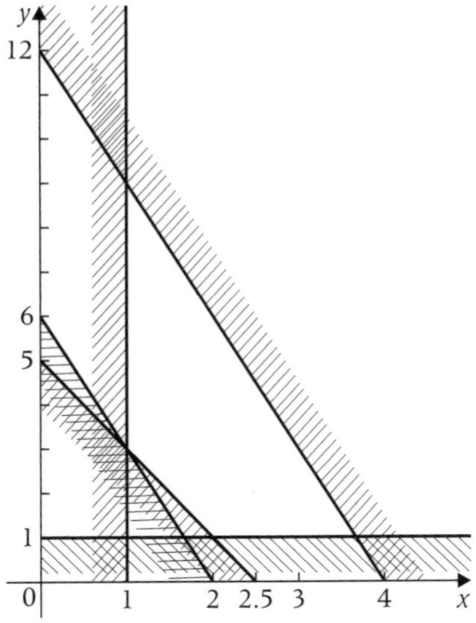

(c) **(i)** Max. (1, 9), $P = 11$

(ii) (1, 8)

5 (a) **(i)** 400 **(ii)** 50

(b) $x \geqslant 0, y \geqslant 0, x + y \geqslant 50, 2x + y \leqslant 200, 0.4x + y \leqslant 120$

6 (a) $x > y + z, 17y \leqslant 3x + 3z, y \geqslant z, 5z < 3x, C = 8x + 12y + 14z$

(b)

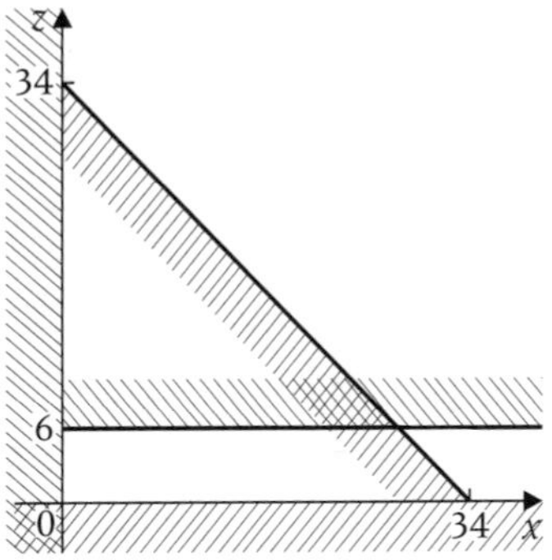

(34, 0), £344 000

7 (a)

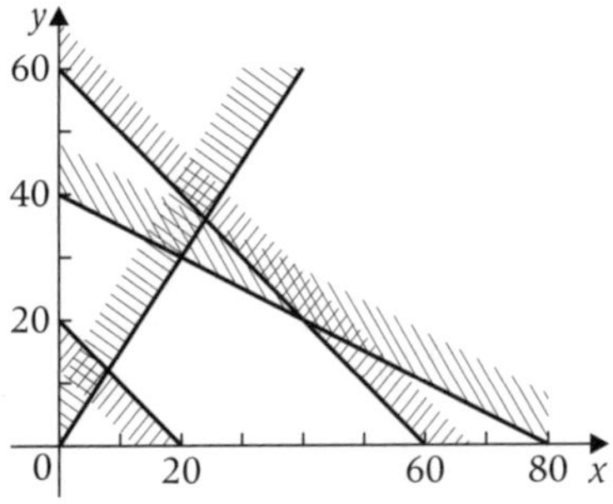

(b) (16, 32), 192

8 $x + 2y + 3z \geqslant 30$, $2x + 3y + 5z \geqslant 50$, $3x + 3y + 4z \geqslant 40$,
$3x + 4y + 6z \leqslant 100$, $z \geqslant 2x + y$

9 **(a)** $x \geqslant 5$, $y \geqslant 20$, $3x + 3.5y \leqslant 210$, $6x + 2y \leqslant 180$, $30x + 40y \geqslant 1200$

(b)

(c) **(i):**

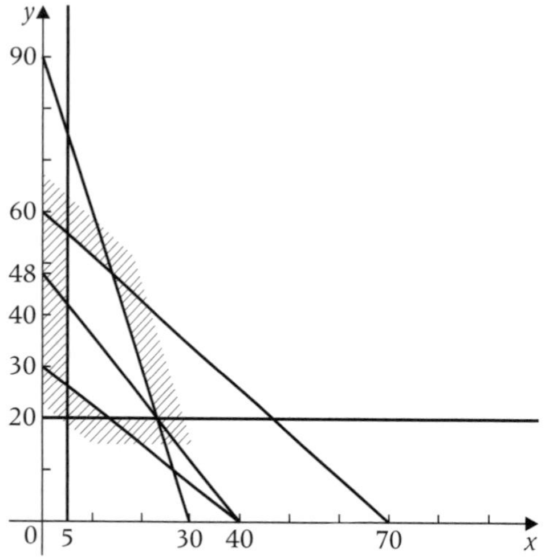

(c) **(i)** $P = 3x + 2.5y$, see graph above

(ii) (14, 48), £162

(d) £233.50

10 **(a)** $1000 - x - y$
$1000 - x - y \leqslant 500$, implies $x + y \geqslant 500$
$1000 - x - y \geqslant 100$, implies $x + y \leqslant 900$
$1000 - x - y \leqslant 2x$, implies $3x + y \geqslant 1000$

(b)

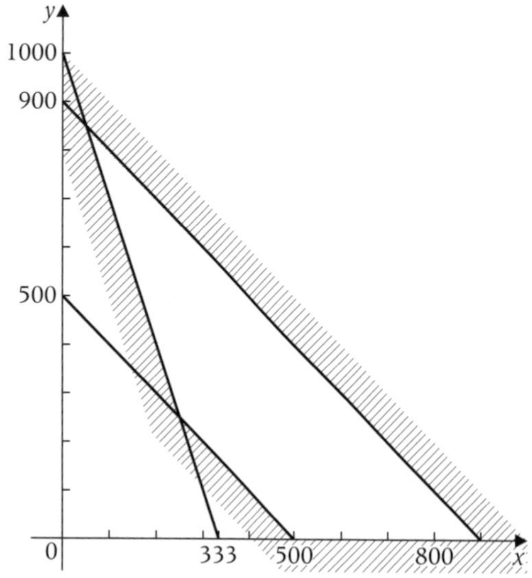

(c) **(i)** $T = 50x + 55y + 25(1000 - x - y) = 25x + 30y + 25\,000$

(ii) (500, 0), $T = £37\,500$
Road 500, rail 0, sea 500

(iii) $y > 500$, implies $R = 38$

11 **(a)** **(i)** $60x + 20y \leqslant 900$

(ii) $x \geqslant 3, y \geqslant 3, 5x + y \geqslant 40$

(b)

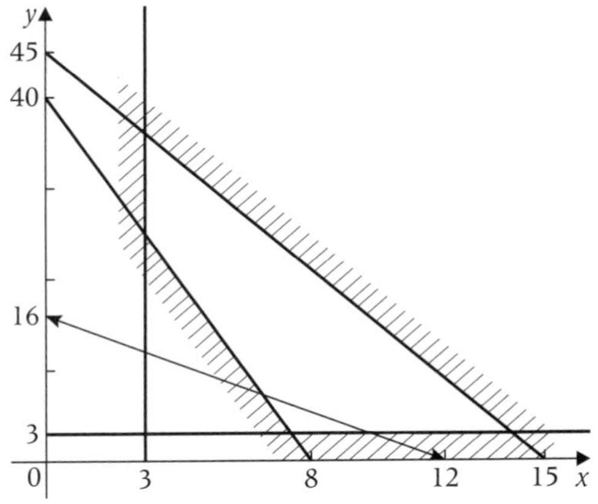

(c) **(i)** (3, 36)

(ii) £64 800

(d) **(i)** (10, 15)

(ii) (14, 3) £1314

12 **(a)** $x + y \geqslant 10$, $12\,500x + 7500y \leqslant 187\,500$ implies $5x + 3y \leqslant 75$, $1000x + 2000y \leqslant 29\,000$ implies $x + 2y \leqslant 29$

(b) **(i)** and **(ii)**

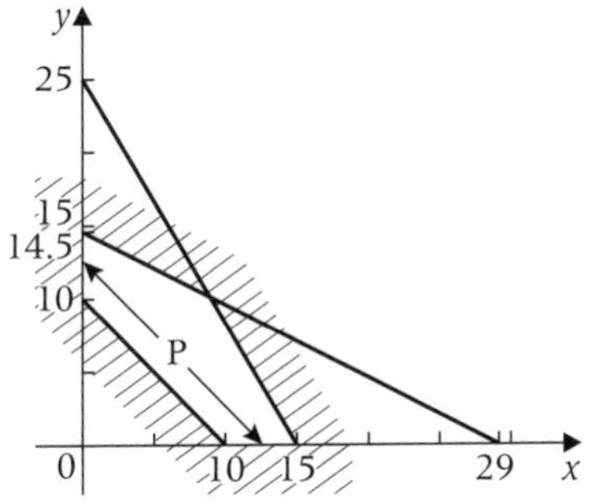

(iii) (9, 10) £22 800

13 **(a)** $90x + 90y \leqslant 1200$ implies $3x + 3y \leqslant 40$, $60x + 120y \leqslant 1500$ implies $x + 2y \leqslant 29$

(b) **(i)** and **(ii)**

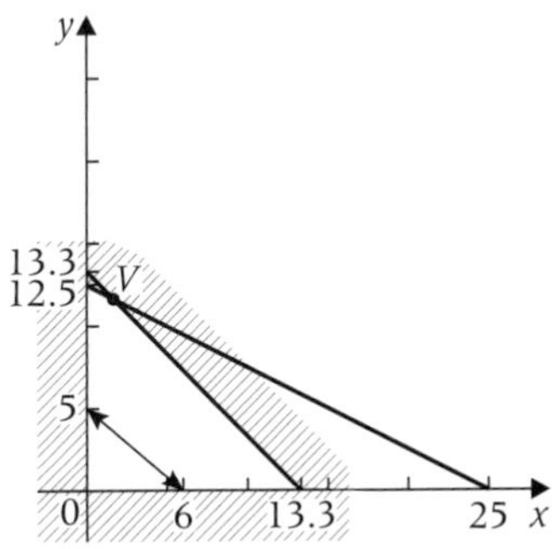

(iii) At V, x and y are non-integers.

(iv) (1, 12), £154

14 (a) $3x + 2y + z \leqslant 120$, $3x + 4y + 6z \leqslant 330$

(b) (i) $z = y$ implies $3x + 2y + y = 3x + 3z \leqslant 120$ or $x + z \leqslant 40$;
$3x + 4y + 6y = 3x + 10z \leqslant 330$.

(ii) and **(iii)**

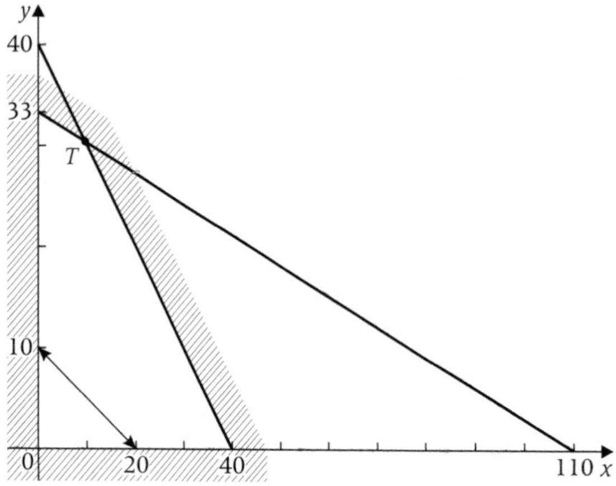

(iv) $(10, 30)$, $T = 70$ **(v)** $(16, 24)$, $T = 64$

15 (a) $50\,(450x + 600y) \leqslant 450\,000$ implies $3x + 4y \leqslant 60$
$100(200x + 150y) \leqslant 300\,0000$ implies $4x + 3y \leqslant 60$

(b)

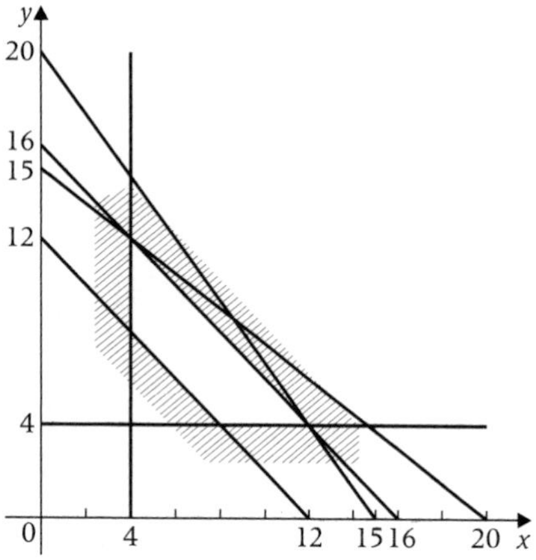

(c) £120 000; (4, 8), (5, 7), (6, 6), (7, 5), (8, 4)

(d) $(4, 12)$, $S =$ £56 250
$(12, 4)$, $S =$ £55 000

Exam style practice paper

1 (a)

A	B	C
1	1	2
1	2	3
2	3	5
3	5	8
5	8	13
8	13	21
13	21	34
21	34	55
34	55	

(b) Generates Fibonnacci series

(c)

A	*B*	*C*
1	1	2
1	2	
1	1	2

i.e. a continuous loop

(d) Add LINE 35 $N = 2$
LINE 55 $N = N + 1$
LINE 56 If $N > 26$ then goto LINE 90
LINE 80 Goto LINE 40

2 (a)

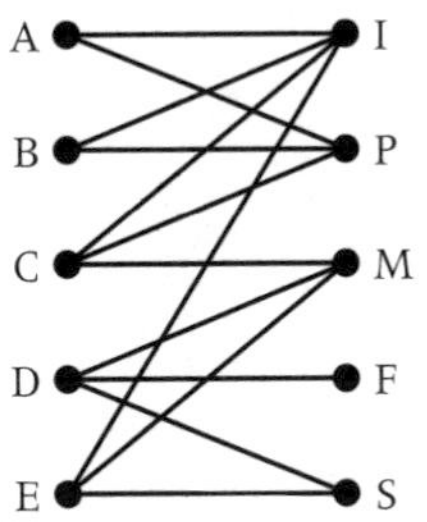

(b) C – M ≠ D – F
AP, BI, CM, DF, ES

(c) D only teacher to teach F.

3

23	2	31	4	30	11	5	13	C	S
2	4	11	5	13	23	31	30	7	5
2	4	11	5	13	23	30	31	5	1
2	4	11	5	13	23	30	31	3	0
2	4	5	11	13	23	30	31	2	1
								17	7

4 (a) Odds A J K L
AJ + KL = 125 + 40 = 165
AK + JL = 105 + 80 = 185
AL + JK = 130 + 40 = 170
Minimum 165 + 550 = 715 m

(b) $7 \times 5 \times 3 = 105$

5 (a)

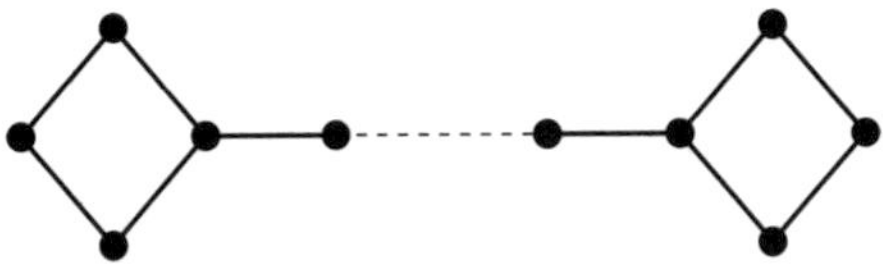

1

(b)

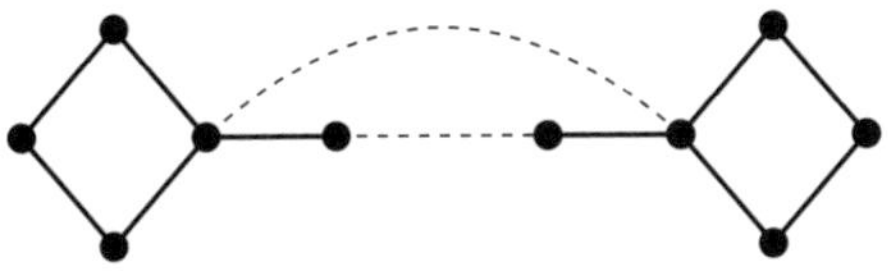

2

(c)

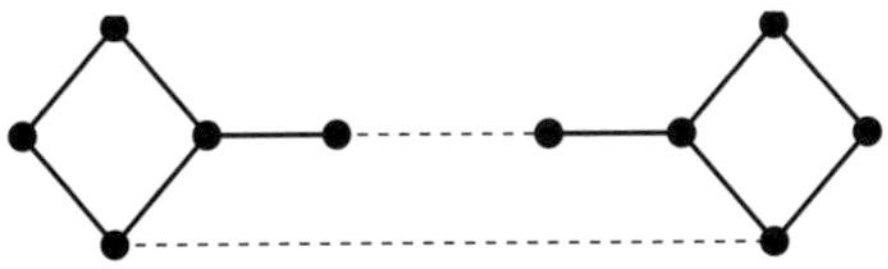

2

6 (a) (i) HM, DE, KH, HI, FE, AB, KL, EA, LN, FG, IJ, BC, AH = 249
9 10 11 11 12 12 13 14 17 20 21 24 75

(ii)

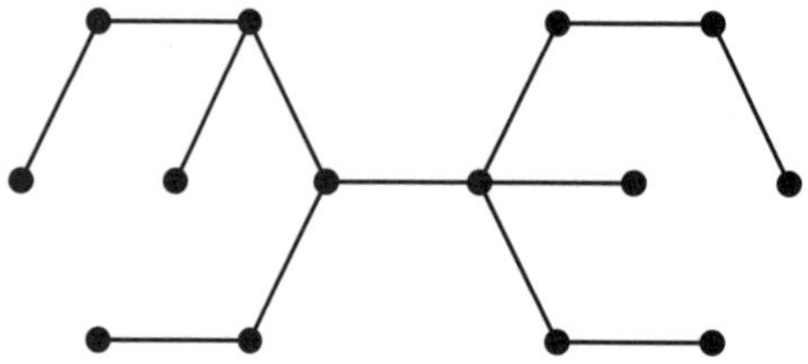

(b) JN = LN = 17 but not IJ
JN = IJ = 21 $x = 21$

7 (a)

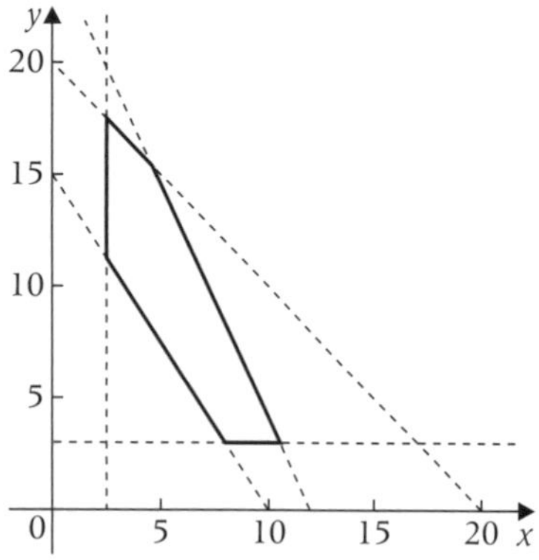

(b) Minimum at (8, 3) = 17
Maximum at (2, 18) = 56

8 (a)

	A	B	C	D	E
A	–	45	80	25	55
B	45	–	35	20	50
C	80	35	–	55	85
D	25	20	55	–	30
E	55	50	85	30	–

(b) A D B C E A = 220
25 20 35 85 55

(c) A D B C B D E D A
B – 2, C – 1, D – 3, E – 1

9 $x + y + z \geqslant 180\,000$
$x \geqslant 20\,000,\ y \geqslant 20\,000,\ z \geqslant 20\,000$
$x \geqslant \frac{40}{100}(x + y + z) \Rightarrow 3x \geqslant 2y + 2z$
$z \leqslant \frac{25}{100}(x + y + z) \Rightarrow 3z \leqslant x + y$
$E = 36x + 32y + 22z$

Index

adjacency matrix 99
Al-Khowarizmi, Mohammed 135
algorithms
  flow diagrams **137**–138
  instructions in pseudo-English **140**–142
  introduction **135**–136
  key point summary 148
  sorting 123–133
  stopping conditions 142–143
alternating path **114**
answers to exercises 181–210
arcs 90

bipartite graph 94, 109–110
bubble sort algorithm **123**–125

Chinese postman problem
  algorithm **48**–49
  Eulerian trail **45**–**46**–47
  finding the route 50–51
  introduction 44
  key point summary 62
  pairing odd vertices 47–48
  semi-Eulerian trail **46**–47
  traversable graphs **45**–**46**–47
  variations 52–53
complete graph **93**–94
complete matching 112
connected graph 91
connected vertices 91
cycle 96

degree of vertex 92
digraph 93
Dijkstra's algorithm
  edge with negative value 32
  finding shortest route **33**–34
  key point summary 41
  limitations **32**–33
  multiple start points 30–**31**
  shortest path problem **25**–29
directed edges 93
directed graph 93
directed networks 25
disconnected graph 91

edges
  definition 90
  directed 93
Eulerian graph 97
Eulerian trail **45**–**46**–47, 97
exam style practice paper 177–180,
    answers 210–212
excluded region 152

feasible region **152**, **158**
finite region 153
flow diagrams **137**–138
FOR loop 142
fully connected graph 91

graph theory
  definitions 90–101
  key point summary 106–107
graphs
  adjacency matrix 99
  bipartite 94, 109–110
  complete **93**–94
  connected 91
  digraph 93
  directed 93
  disconnected 91
  Eulerian 97
  Eulerian trail **45**–**46**–47
  fully connected 91
  inequalities 151–**152**–153
  semi-Eulerian 97
  semi-Eulerian trail **46**–47

simple 91–92
solutions for linear programming 157–**158**–**159**–166
weighted 91
greatest lower bound 72–73

Hamiltonian cycle 96, 104

IF THEN statement 142
incomplete networks 76–77–78
inequalities, graphs 151–**152**–153
instructions in pseudo-English **140**–142

key point summaries
algorithms 148
Chinese postman problem 62
Dijkstra's algorithm 41
graph theory 106–107
Kruskal's algorithm 20
linear programming 175
matchings 121
minimum connectors 20
Prim's algorithm 20
shortest path problem 41
sorting algorithms 133
travelling salesman problem 88
Kruskal's algorithm
comparison with Prim's algorithm 11
key point summary 20
minimum spanning tree **3**–5
Kuan Mei-Ko 44

linear programming
constraints 154
excluded region 152
feasible region **152**, **158**
finite region 153
graphical solutions 157–**158**–**159**–166
graphs of inequalities 151–**152**–153
introduction 150–151
key point summary 175
objective function 154
optimal point 159
optimal value 159
problem formulation 153–**154**–156
solution region 152
three variables 165–166
loops 91
lower bound **72–73**
lowest upper bound 68

matchings
alternating path **114**
complete 112
definition **112**
introduction 109–110
key point summary 121
maximal 113–**114**–117
maximum 112
maximal matching 113–**114**–117
maximum matching 112
minimum connectors
introduction 1–2
key point summary 20
Kruskal's algorithm **3**–5, 11
Prim's algorithm 7–9, 11, **12**–14
minimum spanning tree 2, 98
multiple start points 30–**31**

nearest-neighbour algorithm **68**–69, 71–72
networks
directed 25
incomplete 76–**77**–78
introduction 1–2
triangles 24
weighted graph 91
networks, introductory sections
Chinese postman problem 44
matchings 109–110
minimum connectors 1–2
shortest path problem 23–24
travelling salesman problem 65–67
nodes 90

objective function 154
optimal point 159
optimal value 159
order of vertex 92

pairing odd vertices 47–48
path 95
Prim's algorithm
comparison with Kruskal's algorithm 11
key point summary 20
matrix form **12**–14
minimum spanning tree 7–9
problem formulation 153–**154**–156

quick sort algorithm **129**–130

semi-Eulerian graph 97
semi-Eulerian trail **46**–47, 97
shell sort algorithm **127**–129
shortest path problem
  Dijkstra's algorithm **25**–29
  finding shortest route **33**–34
  introduction 23–24
  key point summary 41
  multiple start points 30–**31**
  triangles in networks 24
shuttle sort algorithm **126**–127
simple graph 91–92
solution region 152
sorting algorithms
  bubble sort **123**–125
  efficiency of methods 131
  introduction 123
  key point summary 133
  quick sort **129**–130
  shell sort **127**–129
  shuttle sort **126**–127
spanning tree 98
stopping conditions 142–143

trail 95
travelling salesman problem
  greatest lower bound 72–73
  Hamiltonian cycle 104
  incomplete networks 76–77–78
  introduction 65–67
  key point summary 88
  limitations of nearest-neighbour algorithm 71–72
  lower bound **72–73**
  lowest upper bound 68
  nearest-neighbour algorithm **68**–69, 71–72
  upper bound **67–68**–69
traversable graphs **45–46**–47
tree 97–98

upper bound **67–68**–69

vertex
  definition 90
  degree 92
  order 92
vertices, connected 91
weighted graph 91